The GARLIC PAPERS

Also by Stanley Crawford

NONFICTION

The River in Winter: New and Selected Essays

A Garlic Testament: Seasons on a Small New Mexico Farm

Mayordomo: Chronicle of an Acequia in Northern New Mexico

FICTION

Village: a novel

Intimacy

Seed

The Canyon

Petroleum Man

Some Instructions to My Wife Concerning the Upkeep of the House and Marriage and to my Son and Daughter Considering the Conduct of their Childhood

Log of the S.S. The Mrs. Unguentine

Travel Notes

GASCOYNE

The GARLIC PAPERS

A SMALL GARLIC FARM IN THE AGE OF GLOBAL VAMPIRES

STANLEY CRAWFORD

LeafStormPress
SANTA FE, NEW MEXICO

The Garlic Papers: A Small Garlic Farm in the Age of Global Vampires

Published by
Leaf Storm Press
Post Office Box 4670
Santa Fe, New Mexico 87502
USA
leafstormpress.com

Book Design by LSP Graphics
Art & Design Consultant: Liza Doyle

First Edition
Printed in the US

10 9 8 7 6 5 4 3 2

Library of Congress Control Number 2019940901
Cataloging-in-Publication Data is on file

ISBN 978-1-945652-05-9

For Ted and Renate Hume
and Thomas Conboy

Work, because it is productive, produces in man a productive hope.

—JOHN BERGER

When danger approaches, sing to it.

—*ARAB PROVERB*

Jarndyce and Jarndyce drones on. This scarecrow of a suit has, in course of time, become so complicated, that no man alive knows what it means. The parties to it understand it least; but it has been observed that no two Chancery lawyers can talk about it for five minutes, without coming to a total disagreement as to all the premises. Innumerable children have been born into the cause; innumerable young people have married into it; innumerable old people have died out of it. Scores of persons have deliriously found themselves made parties in Jarndyce and Jarndyce without knowing how or why; whole families have inherited legendary hatreds with the suit. The little plaintiff or defendant, who was promised a new rocking-horse when Jarndyce and Jarndyce should be settled, has grown up, possessed himself of a real horse, and trotted away into the other world. Fair wards of court have faded into mothers and grandmothers; a long procession of Chancellors has come in and gone out.

—CHARLES DICKENS, *BLEAK HOUSE*

CONTENTS

Chapter 1

TWENTY-FIVE TONS OF GARLIC

The first garlic harvest of the season is typically a small one, a matter of a few bulbs for the host of a dinner party or a visiting friend. I select a half dozen plants to dig with a narrow-bladed shovel, ones whose leaves have yellowed slightly more than the others and whose bottom wrapper leaves have turned brown and shriveled. I insert the blade vertically into the soil a few inches from each plant and thrust it deep into the ground with a sandaled foot. Then I pry back, leveraging the bulbs out of the earth. Although I have done this thousands of times in various ways, I'm delighted at seeing the white skin of the first bulbs peeking out from clods of dark soil and at the aromas of damp earth and freshly exposed garlic that rise into the air. I brush off the clods of dirt clinging to the long white roots, which dangle like jellyfish tentacles from plump bulbs streaked with purple.

The earliest variety is Turban Chengdu, one of four cultivars we grow on our two-acre farm in the village of Dixon, in Northern New Mexico. It usually matures in the first week in

June, two or three weeks before most other cultivars. Its earliness gives us a brief sales advantage at the Santa Fe Farmers' Market. While digging out the first bulbs with a short-handled planting shovel, I invariably recall our first years of harvesting garlic in the mid-1970s. Armed with shovels, garden forks, buckets, a wheelbarrow, along with a dozen city friends and country neighbors, we attacked the garlic patch by hand. One young woman, friend of a friend, arrived in knee-high red leather boots and exclaimed, "I've never been in a field before!" Our first cultivar was a local hardneck, or topsetting. We dug up the whole plant after snipping off the seed tops—or rather, clusters of bulbils—hands and forearms splashed with pungent garlic juice, which whitened as it dried. Years later we discovered that cutting off the top sets just before harvest disrupts the hydraulic economy of the plant, resulting in bulb shrinkage that causes the outer layers of skin to flake off too readily.

Within a couple of seasons, our laborious hand-digging system was replaced by our first tractor pulling a chisel plough down the middle of the garlic beds to loosen the soil around the bulbs. That done, I would hook up a small forklift to the tractor to move the harvest cages into the field. These were homemade wood-and-chicken-wire constructions six feet wide, two feet deep, and three feet tall, open on the back side. Each would carry around 150 pounds of garlic plants up to the storage sheds.

After ten years of this, I moved up to a 30-horsepower four-wheel-drive Kubota, powerful enough to pull two Spin Sweeps, 20-inch disks on rotating shafts, with which to undercut the garlic. The plants are then easily pulled or lifted. Under certain conditions—soft ground, few large weeds—I can also use a Spanish-made undercutter with a conveyor belt that is supposed to lift and deposit the bulbs on the ground, though it sometimes reburies them. A new generation of harvest bins made of plywood and about a foot larger in all directions has since replaced the old cages.

I have grown garlic about thirty-two of my forty years of farming, for a total of around 50,000 pounds or 25 tons. This on three separate fields in the Embudo Valley of Northern New Mexico, north of Santa Fe and south of Taos, at an elevation of around 6,000 feet.

Chapter 2

THE CHINESE GARLIC CONNECTION

In the spring of 2016, my first small harvest was for a specific purpose. The time was a bright sunny morning a few days before the Memorial Day weekend. On the following Friday I would slide the garlic across a conference table to the Deputy Assistant Secretary of Commerce of the United States Department of Commerce in Washington, DC.

After digging up six bulbs of Turban Chengdu, I carried them up to the washing stand under the weeping willow, hosed them down, and left them on the table to dry. Later in the day, I bagged them up in two layers of plastic so they wouldn't stink up my clothes in the suitcase. I added a copy of my book *A Garlic Testament: Seasons on a Small New Mexico Farm* (University of New Mexico Press, 1992). The object of these gifts was to help convince the Deputy Assistant Secretary and his staff that I was a bona fide garlic producer and therefore qualified to ask that the largest importer of Chinese garlic be reviewed in order to establish a new rate of duty.

Now why would you want to do something like that? you may well ask. The answer is long and complicated, but a short version is this: US law prohibits foreign producers of goods from dumping. Dumping means selling their product at an unfairly low price, thereby undercutting domestic producers of the same goods. Foreign companies that are found to be dumping their products into the US market are assessed a duty, or tax on the goods they're selling. As of 2018, the largest importer of Chinese garlic had gotten away with paying zero duty for the past fourteen years, with depressing effect on US garlic prices and domestic garlic production.

Once a year, any domestic garlic producer can request the US Department of Commerce to review any or all importers of garlic. With the assistance of an experienced trade attorney, I requested a review in the fall of 2014. At the time neither he nor I realized the size of the hornets' nest we were poking at. Neither of us could have foreseen that we would find ourselves sued by the world's largest importer of Chinese garlic, vilified in public forums, and betrayed by a neighboring garlic grower. My one-acre patch of garlic was about to become a pivot around which a dozen international and local law firms would circle, representing individual and corporate petitioners, plaintiffs, and defendants, in two countries and four US legal and administrative forums.

Chapter 3

CLONES BEGET CLONES

I suppose it's appropriate that Turban Chengdu, which presumably originated in the Chengdu region of China, is the cultivar I took to Washington in connection with Chinese imports in May of 2016. Turban Chengdu is a recent addition to our line of cultivars, which includes Russian Red, Siberian, and Bosque Early. The latter started out as California Early, a hundred pounds of which was originally purchased from Christopher Ranch in Gilroy, California. There is some irony here, as will become apparent later.

Back in the 1970s, when we first started growing garlic, discussions were confined to California Early, a white softneck garlic, and California Late, a softneck with some purple streaking on the outer skin. We started not by growing either of these but rather with a nameless local cultivar often found in the semi-wild in apple orchards of our narrow river valley. Garlic was among the Eurasian cuttings, seeds, roots, and bulbs brought to New Mexico by Spanish colonists in the sixteenth and seventeenth centuries. The topsetting garlic growing here and there in the Embudo Valley could well have descended from those early plantings. But we have no way of knowing.

Russian Red I obtained from a grower in Albuquerque, who could tell me where he got it from, but I doubt there is a verifiable chain that goes all the way back to Russia. The same can be said of the Siberian, obtained from a grower who lives on the northern New Mexico border. So it probably goes for the dozen other cultivars sold at the Farmers' Market in Santa Fe and the hundreds more listed in seed catalogs and the Seed Savers Exchange, many with designations including the names "Spanish," "German," "French," Romanian," "Vietnamese," and so on. The best chain of provenance I have is with Bosque Early, named for our farm, El Bosque. *Bosque* is the New Mexican Spanish term for the growth of willows, cottonwoods, New Mexico olive, and squawberry bush along riparian corridors. Bosque Early is my version of California Early, which has been grown in the Central Valley of California for decades, if not longer. But it too is descended from Eurasian cultivars.

Garlic cultivars do not reproduce sexually but clone instead. The variations among them are called land races, which are cultivars that have adapted over time to the soil and climate conditions, including day length, where they are grown for long periods of time. Of the four cultivars we grow, Turban Chengdu has not yet settled down to our climate. We've grown it now for five years. During the first years, a high percentage of bulbs sent up seed stalks, which we continue to cull out. The cultivar also generated a high proportion of large "rounds," garlic bulbs without clove divisions. For uniformity, we have planted these separately from the others. The rounds will yield large bulbs with clove divisions the following season. Eventually I hope the cultivar will adapt to our conditions and that I can continue to increase production.

We stopped growing our original Bosque Topsetting after a number of years, when the bulbs became smaller and smaller, probably because we inadvertently selected the smallest bulbs for replanting. Then we switched to Music, a white topsetting cultivar, but it turned out to be too mild for most tastes. Since then, Russian Red has proven reliably good, with large, strongly-flavored cloves. The verdict is still out with Siberian, which may bestill adapting to our somewhat lower altitude and warmer climate.

Chapter 4

GUESTS COME TO THE TOWER WITH A MODEST PROPOSAL

We grow our garlic on three small fields, the smallest of which, an eighth of an acre, we call the Tower field, so named because it is bordered on the south by a driveway spur that leads to our guest house, the Tower. The Tower started out as a one-story circular stone structure, thirteen feet in diameter, built by my wife, RoseMary, and me in the summer of 1972 as my first writing studio. It was more a pillbox than a tower, but I had in mind the defensive torreones built by the first Spanish settlers as refuges against Comanches and Utes in Northern New Mexico. The ruins of one such tower still stand in the old plaza of Dixon. In the back of my mind was also the Martello Tower in Sandycove, Ireland, where young James Joyce lived briefly around 1900, not far down from Killiney Hill, south of Dublin. We spent eight months there in 1968; it's where our son Adam was born. A further echo extends back to William Butler Yeats's collection of poems entitled *The Tower* (1928).

For the first two rooms of our house built the summer before, we had the help of many friends making adobe bricks, pouring the foundation, and laying the roof beams and planks. For the writing studio project, however, I wanted no outside help. This was to be my private space a hundred feet south of the house, round, cave-like, with small windows.

RoseMary and I spent much of the late spring and early summer of 1972 driving our old 1947 Chevy one-ton flatbed up onto the basalt-strewn mesa west and north of our narrow river valley to gather stones for the future structure. As we crept up the rough track to Horserace Mesa, as the locals call it, the Chevy's high clearance and large 17-inch wheels maneuvered the ruts and rocks as well as any four-wheel drive. Bed laden with a mound of the dark volcanic rocks, we descended in compound low gear back down to the highway. The old truck had a crash box, which meant that you had to double-clutch every time you shifted. Bumps often unlatched doors. The front axle kingpins were badly worn, posing a challenge to steering, especially when heavily laden. My children recently reminded me that I was in the habit of rolling cigarettes with Bugler or Bull Durham tobacco while driving, a fact I had conveniently forgotten.

The old Chevy built the house—and the Tower. We would fire up the gas motor of a borrowed cement mixer, hose down the stones still in the bed of the truck, and one by one fit them into the foot-wide bed of mortar, laying up perhaps a course a day around the three window boxes and the low door frame. For the roof beams, we used ponderosa logs eight inches in diameter that we'd gathered ourselves from the forests of the Sangre de Cristo Range twenty miles and a couple thousand feet above our valley. We planked these over with rough-cut boards, then laid down tar paper and black plastic sheeting, on top of which we spread a four-inch layer of dirt. Tradition called for a foot-deep layer, but our beams wouldn't hold that. Happily, the roof leaked only around the stovepipe. The interior diameter of the space was eleven feet.

The Tower was to be a writer's room of my own. (I had recently discovered the work of Virginia Woolf). Time to get another novel out. The last of the movie money from my first novel—written in the idyllic village of Molyvos, on the island of Lesbos—paid for the land, though no movie has ever been made. RoseMary would look after the two toddlers up at the house during my writing mornings. At the time, the world we lived in still considered this an acceptable arrangement. The Tower would be a perfect writing space.

Wrong. It was too small. There was hardly room to pace during those moments when physical action offered the possibility of limbering up the imagination as well. The Tower was either too hot from the wood stove or too cold when the fire died down. It was too far to go up to the house in the snow for coffee, snacks, or to the bathroom. And every time I stepped out the low door, the white Chinese weeder geese in the nearby pen heckled me, cackling (or so I imagined), "Hey, when are you going to write that best seller, huh?"

I stuck it out for two writing winters. I replaced the Chinese weeders with placid gray Toulouse. Eventually I moved back to a desk behind the front door of the living room—a location in her house where Jane Austen wrote. As I typed away each morning, my two toddlers became Head Starters, kindergartners, elementary schoolers, then high schoolers, whereupon in succession they discovered the unused Tower as a way to leave home without actually leaving home. RoseMary's younger sister from Australia, Zita, and her young son, Ruben, lived there a couple of summers in the 1980s. Otherwise it had stood empty.

Then, around 2007, we decided to add on to the stone structure and turn it into a paid guest house, a bed and breakfast but with breakfast consisting only of some yogurt and half-and-half and orange juice left in the fridge, plus coffee and basic condiments. A retired architect friend, Ron Rinker, came up with a simple plan to add a second story in a way that enclosed only a fourth of the attractive circular basalt wall. The old writing space became the bedroom. Across a hallway, the new adobe construction enclosed a bathroom. The kitchen-living area was upstairs, accessible by a spiral staircase.

The roof of the old circular stone structure became a deck overlooking the three fields of our small farm and, beyond, the willows and cottonwoods lining the Río Embudo, which runs along the south side of our dead-end lane. From the deck you can also glimpse the southernmost spur of the Rockies, the Sangre de Cristo Range, fifteen miles to the east. Through the cottonwood trees two miles away lies the basalt mesa that marks the westernmost boundary of the Embudo Valley. And to the north, a view of rocky piñon- and juniper-spotted hillside above the irrigation ditch and our long adobe house with its attached greenhouse. The new design was so attractive that RoseMary and I wanted to move into the Tower ourselves, abandoning a life-long accumulation of household goods and clothes and books.

We soon adopted the custom of inviting paying guests up to the main house in the evening for coffee, tea, or a glass of wine.

After all, they had chosen this out-of-the-way spot without the benefit of tourist brochures or maps, only the brief descriptions on the websites we advertised on. For one reason or another they picked a working farm on a dead-end lane a mile and a half from the center of a traditional Hispanic village. We were curious to know who our guests were, what they did or had done for a living, where they were from, how they had found out about the guest house, and why they had chosen to stay here. Over the years we have hosted a succession of teachers, doctors, lawyers, builders, engineers, mechanics, computer specialists, writers, artists, musicians, members of the military, pilots, students, their children and dogs, and occasionally a cat. Over time a number of guests have become friends.

Ted and Renate Hume booked a couple of nights in the Tower in late October 2014. We invited them up for a glass of wine. Ted, in his late sixties, was a warm and personable fellow, with a generous open face and smile. His wife, Renate, an attractive, well-dressed woman, also warm and outgoing, was a painter and graphic artist. I later learned that the elegant silver jewelry she wore was her own creation. At the time they were living in Ojai, California, in the hills above Santa Barbara. German-born Renate came to the States when she was thirty and lived in Taos, 25 miles north of the farm, for eight years in the 1990s. I immediately liked and trusted both of them, as did RoseMary and her sister Zita, who was staying with us for a month while I went off to teach Southwest Literature at the Center for Southwest Studies at Colorado College, in Colorado Springs. RoseMary's increasing memory loss had been recently diagnosed as Alzheimer's, and she could no longer be left alone for more than short periods of time, though she could still be warm and sparkling with guests. Thomas Conboy was also with us that evening, by now a long-time multitalented farm worker and household helper and friend of the family.

It turned out that Ted and Renate had a purpose in staying at the farm. As an international trade attorney with forty years' experience, Ted was seeking a garlic farm to challenge the largest importer of Chinese garlic, which had legally got away with paying zero duty for the past ten years. This importer, Harmoni International Spice (US), is owned by a Chinese exporter, Zhengzhou Harmoni Spice (China). The other Chinese importers Ted represented had been locked out of the US garlic market for several years on account of being unable to compete with Harmoni's duty-free prices. All importers except Harmoni had been subject to detailed Commerce reviews and had been assessed various rates of anti-dumping duty. Ted explained that as a domestic garlic producer, I had the right to protest this situation with the United States Department of Commerce. He had tried without success to find a California garlic grower to submit a request for review, but no one there would take on Christopher Ranch, America's largest garlic distributor, which presumably benefited from reselling cheap Chinese garlic.

Might I be interested? "No risk, no cost," Ted assured us all.

"Sure," I said naively. "Sounds interesting," I added, with no inkling of how very interesting and how incredibly complicated it would become over the next three years—so interesting, in fact, that it would become the subject of a Netflix documentary show.

Nor did Ted, for that matter. "Unprecedented," he was later to say with some frequency, in response to some surprising turn of events among the trials and tribulations that have bonded us into a deep and lasting friendship. "In my forty years of dealing with international trade law, this is unprecedented."

Chapter 5

CLOAK AND DAGGER COME TO THE FARM

Within weeks after Ted and Renate Hume stayed in the Tower, I had signed a form enabling Ted to represent me in filings with the Department of Commerce. Our hope was that the review Ted and I were seeking would result in a new duty rate. Commerce determined that garlic was among the three hundred fifty products that were being dumped into the American market at prices below cost. Harmoni Spice had paid no anti-dumping duty for over a decade thanks to a loophole in the anti-dumping statutes. The duty can be as high as $4.71 a kilo. A new duty rate would likely bring prices of imported Chinese garlic more in line with domestic prices. It would also level the playing field for smaller Chinese exporters of garlic.

After a month teaching at Colorado College, I returned home to find a well-dressed middle-aged fellow standing in the drive, wanting to talk to me about growing garlic in Texas. He said his name was Mr. Perez. Short black hair, black vest, new black Lexus SUV. He didn't quite fit what I have come to see as a stereotypical bureaucrat nursing dreams of getting back to the land and living out his retirement days in a bucolic setting, but I was untroubled by doubts or suspicions.

The day was December 21, 2014. I had left Colorado Springs by car just before noon, right after my last class session. I was a little tired from the drive and anxious to unpack, but I stood in the driveway and answered his questions for a little while. Mr. Perez asked about the farm and about growing garlic, taking notes as I talked. I said something to the effect that he wouldn't make money growing garlic, but that farming had given me a good life. He didn't ask to see any of my equipment.

Mr. Perez returned the afternoon of January 14, 2015, clipboard again in hand, with some follow-up questions. I was just back from a day in Santa Fe and was impatient to get on with my life. A male companion sat in Mr. Perez's Lexus throughout our conversation.

A month later, while again in Santa Fe on a shopping trip, I got a call from Ted's office manager in Ojai, California, where Ted and Renate had been living for the past ten years. "Are you aware that a private investigator visited your farm—not once but twice?"

"What?" I said. "No way, absolutely no way."

The purpose of his two visits to the farm was not at all clear to me until Ted's Ojai office emailed me a copy of the seventy-page Department of Commerce filing against us. This was when I learned that my prospective grower of garlic in Texas was in reality private investigator Bernardo Perez, working for none other than Harmoni Spice, whose zero duty rate we were challenging with Commerce. According to Perez's resumé, submitted as part of their filing against my farm, he had worked extensively for the FBI in this country and in Latin America.

Harmoni was opposing me on its own behalf and on behalf of Christopher Ranch and the three other members of the California-based Fresh Garlic Producers Association. The filing attempted to undermine my "standing" as a commercial producer of garlic and thereby disqualify me from requesting a review of Harmoni's business practices, including production and shipping costs.

Harmoni was represented in its administrative filings with the Department of Commerce by the massive international law firm Grunfeld, Desiderio, Lebowitz, Silverman & Klestadt LLP (GDLSK), with offices in New York, DC, Los Angeles, and Hong Kong.

I was amused by the attention both these goliaths, Harmoni Spice and GDLSK, were paying my little operation. I was also somewhat unnerved: what might they be planning next?

Chapter 6

ΚΑΤΑΣΚΟΠΟΣ

In his written statement to Commerce filed by Harmoni's lawyers, Mr. Perez claimed he only saw "an old tractor." True, the Kubota L2850 parked in the shed was old, almost thirty years old, but "old" is not a disqualifying factor in the case of farm equipment, which can serve for three or more generations. But this is a quibble. What he did not see or note in his report was the tractor-mounted rototiller (forty years old) and furrower and brush hog and bed shaper and blade and box scraper and fork lift, down between the Tower and the chicken house, under the apple trees. Nor did he notice the Holland transplanter down there, our first garlic planter, nor its replacement behind the house, a one-row carousel-type Lannan transplanter from Finland. Nor did he ask to poke around in the garlic sheds, where he would have seen the motorized brusher-cleaner we use to clean garlic and shallots, or the hundred or so antique Palmer apple boxes we use to store and transport garlic and shallots and other crops. Nor the huge stack of long one-bys and short two-bys with which we construct some seventy feet of drying racks for garlic and shallots each summer. Down by the gate, he drove by four times without noticing the undercutting disks on a toolbar and three large plywood harvest bins we use to move garlic and other crops from the field to the sheds. And of course he did not see inside my studio, where are stored boxes of financial and tax and depreciation and employment records for the farm. Nor could he know that virtually our entire acreage was planted in garlic and shallots, whose bulbs and cloves had begun to root but would not send up any green shoots until late February or early March.

Following his two deceptive visits to the farm, friends and family expressed concerns about my physical safety, which subsequent events did nothing to allay. At first I was more amused that Harmoni would go to the trouble to hire a private investigator to visit my small operation and poke around public records (fortunately alerting me that I had forgotten to apply for a county business license for the current year) than I was fearful that it might lead to something more serious. The a dvantage of living in a small village is that everyone knows you—and you know everyone. Yet I became more watchful for unfamiliar cars on our dead-end lane. At one point I wondered whether a black Cadillac Escalade, parked nose out at the post office, was there for a sinister purpose.

I have always been fond of the Greek word for spy: *kataskopos*, literally meaning someone who looks beneath the surface. As a spy, as a private investigator, Mr. Perez was not very good. I wouldn't recommend him.

I would also call him a liar, except, who knows, perhaps he has retired from spying on small farmers for a large international corporation and has indeed moved on to growing garlic in Texas. In which case, I wish him well. He will probably not make much money, but he will find it a good life, and certainly an ethical improvement over his previous employment.

Chapter 7

DAS KAPITAL

Had he inquired, I would have told Mr. Perez how capital-intensive even small farms are. Over the course of forty-some years, I have invested in succession in four tractors, four two-wheel rototillers, three transplanters, a flail mower and manure spreader, three storage sheds, a walk-in cooler, a dozen pickups and vans, and various drip irrigation systems, not to speak of land, both bought and rented.

In an area like Northern New Mexico, where the relatively small amount of irrigated land is subject to speculation pressures, a young family in 2019 would need $500,000 or more to set themselves up on five acres with the proper equipment, adequate housing, and modest storage space. Few can start with this kind of stake. Many, like RoseMary and myself, nickeled and dimed our way up the shaky economic ladder. In local terms, we paid too much for our two acres of brush and apple trees. In 1970, $2,300 was considered high, but it was enough to buy a new Dodge Swinger, which is what the former owners used the money for. Now the land alone is worth fifty times what we paid. We built most of our house ourselves without the exacting requirements of building inspectors—and legally, because in New Mexico, farms are exempt from building codes.

Our first tractor was a derelict Farmall A rescued from the postmaster's barn and revived. It was owned collectively by three of us. My first truck, the twenty-year-old 1947 Chevy one-ton, cost very little and was unusually reliable in a falling-apart way. The easygoing bank managers of the former Valley National Bank in Española lent us money to buy two more acres, which qualified us for the first of two USDA Farmers Home Administration (now the Farm Service Agency) loans, with which we bought tractors: two in 1977 and three in 1986. Family loans also kept us going, plus a Valley National Bank manager, Lillian Martinez, who approved yearly operating loans year after year.

We were lucky. We bought land when it was still affordable. A number of young and even middle-aged farmers in Northern New Mexico currently don't own land. They are at the mercy of "handshake leases," which is to say at the mercy of landlords who don't recognize their tenants' investment in the sweat equity that keeps the land productive, maintains water rights, repairs ditches, fences, and gates. As one young woman farmer once told me, "I won't plant trees." Not because they wouldn't grow, but because she's uncertain that she'll be still using the same land in the distant future.

There are financially successful market farmers in Northern New Mexico, but I would guess that the younger farmers are just squeaking by. They have no retirement plans, no child care, sick leave, or college funds for their growing children, and they may not have health insurance. But for those who can live with this degree of economic marginalization, there are benefits. You can work outdoors at hours of your own choosing, at least when crops are not dictating your schedule. Long winter vacations in the cheaper parts of Mexico are possible for some. And overall you have a high degree of independence from your fellows. Plus the satisfaction of knowing that some of your food, or most of it, or even all of it, comes from the labor of your own hands and those of your partner.

In an era of increasingly complex forms of internet-enabled dependencies, this is no small thing. A working farm becomes the center of the universe for those who work its land and crops and tools and machinery, a refuge and a sanctuary from the gravitational forces of the twenty-four-hour news cycle and social media trying to pull you away from a place-centered life, luring you to click your way out into the void.

Chapter 8

LIVE POOR, DIE RICH

Regarding indebtedness, a couple of thoughts. As a farmer, you have to weigh financial independence against the allure of equipment and devices and other kinds of improvements that will save labor and make you more productive, and that promise to gain you some advantage in the race of bodily vigor against the grinding effects of time. The risk of frugality is that you end up too worn out physically to enjoy the labor-saving devices that would have forestalled your aches and pains. I used this argument to justify the purchase of tractor number four, a 30-horsepower Kioti with a semiautomatic transmission that required far less left-foot clutch work than the old Kubota. Better that than a hip replacement.

There is of course a dark side to indebtedness. As a debtor through the mainstream banking and credit card system, you are transferring wealth upward: your interest fuels investments in industries and causes likely at odds with your beliefs and practices. This was less the case with the two USDA loans that helped us upgrade our equipment and increase production, and not the case with the Santa Fe Farmers' Market Institute Microloans Program, from which I have borrowed to buy a 20-by-30-foot hoop house, to rebuild and improve our 14-by-36-foot solar greenhouse attached to the house, and to buy a four-row Polish garlic planter, which has radically reduced planting time for garlic and shallots. The Micro Lending Program has also temporarily financed irrigation improvements, including a ten-thousand-gallon storage tank, most of whose costs will be eventually covered by a grant from the USDA Natural Resources Conservation Service. Minus processing costs of the credit union that administers the Microloans Program, loan interest revolves back into the fund, serving to grow it and help other farmers.

Given that we work with and often against the weather, and with plants and animals and insects and the soil itself, it is no wonder that contingencies and emergencies are often the rule. I find it difficult to remember the days before credit cards, which enable spur-of-the-moment purchases for which farming offers too-easy excuses. I have been constantly urged to budget, but how can you budget against the weather in the era of rapid climate change? Or against technological change that can exact premium prices in a world where food prices remain stuck in the 1990s? I pay more now for information services than I do for gas and diesel and electricity combined.

In this disparity between rapid technological change and low commodity prices lies the widening gap between rich and poor and the yawning maw of credit-card and student debt.

As a farmer, the long shadow of debt is likely to loom over your fields until it doesn't matter anymore, until debt relief finally comes in the form of release from all worldly cares and joys.

With luck, your heirs will enjoy the fruits of your sweat equity, and what is left over from paying off your debts.

Chapter 9

"EL BOSQUE GARLIC FARM WALKS INTO A MCDONALD'S . . ."

When Harmoni filed against me, they claimed I was only a hobby farmer, lacked a business license and other documentation, made no money selling garlic. Their seventy-page filing included this memorable line: "El Bosque Garlic Farm is such a fiction that it couldn't walk into a McDonald's and order a cheeseburger." It's hard to know what to call this figure of speech. Or disfigure of speech. The line was probably written by a young law intern trying to show some literary flair.

As a writer of fiction, I know well that none of my fictional creations are mobile or hungry. They are thoroughly inert in book form until bathed in the solvents of consciousness and the imagination. At the other end of the spectrum, as a writer of nonfiction, I have recorded countless observations about farming over the decades. To most observers, other than our young law intern, my writings have had the effect of convincing the reader of the reality of my farm.

The cheeseburger statement reveals Harmoni's intentions to turn my farm—and therefore a good part of my life—into a legal fiction in order to induce the Department of Commerce to dismiss my administrative review request.

Harmoni's lawyers were probably not aware that turning things into fiction, though not for legal purposes, was how I started out my professional life as a novelist in 1964. I inherited the artistic gene inherited from my mother—though my father was a better writer, to judge from his spare but witty letters. Mother: shy, hesitant, with the somewhat halting speech of someone whose first language was not English. She was the daughter of a German couple who immigrated to the States around 1910. Until her fifties, she wore her hair in the German style, in double braids fastened across the top of her head, parted down the middle. After high school in Los Angeles, she somehow summoned up the courage to enroll in the UCLA arts program in the late 1920s, obtaining a degree in painting. Unfortunately, she married into a family of philistines, the Crawfords (Dad's mother was a Bergman from Illinois German stock), and spent the next twenty years raising my sister Diane and myself. When I was in elementary school she became a maniacal producer of hooked rugs, cutting woolen material into narrow strips and dyeing the strips on the stove in large pots, then hooking the strips into the flowery patterns mapped out on burlap backing.

When I was in high school, Mother began talking of taking painting classes again. I thought I should encourage her to act on her desires: I gave her an easel for Christmas. When she opened the large box, she burst into tears. I was perhaps the first in the family to acknowledge her suppressed aspirations, at least in a tangible way. And when I went off to college, she finally enrolled in painting and drawing classes. Many years later, in the mid-1980s, she won a prize for a pastel of our adobe house and the flower garden in front of it, which now hangs in our New Mexico living room.

But then, a year after she'd completed the drawing, a call came from Dad. Mother was in the hospital being treated for a stroke that had rendered her sightless. I had to fly home to San Diego the next day. We had just finished laying out our garlic to dry in the driveway. I couldn't leave it exposed to the July rains for the three or four days I would be gone. It was then that I assembled the first drying racks in the shed, made from a pile of scrap wood left over from various construction projects.

During my undergraduate years at the University of Chicago, I wanted to paint more than write. A crucial turning point came when I had to choose a major at the end of my second year. I checked out several departments. Art history was too dry and academic. The painting department occupied an old house on the south side of the Midway. It was presided over by a friendly older professor-painter. I was very drawn to the program, but unlike my mother I lacked the courage to take the leap. I enrolled instead in the safer English Department.

Yet, the urge held on. I signed up for Saturday painting classes, first at the Art Institute of Chicago and then several years later at the San Francisco Art Institute. In between the two, in Paris as a student, and then in Cali, Colombia, as a teacher of English as a second language, I dabbled at painting. Or more than dabbled in Cali, where I spent weekends painting with a fellow teacher's wife in their airy high-rise apartment. When we both accumulated enough work, we mounted a show at La Tertulia Gallery in downtown Cali. Nothing sold.

In Berkeley as a grad student in English, seeking to fill out my reading of the classics, I came to realize that I was too shy and private to be a painter. And I would have been too shy to take creative writing classes, had there been any available at Chicago or Berkeley at the time. I was then in a phase of depicting on canvas my current confusions of sexual identity in the form of young men who were themselves agonizing. Was I gay or bi or what? If I wrote instead of painted, no one would see what I was writing, and thinking, and feeling, until I was ready to reveal my words. In Berkeley I typed out my one and only short story. I never submitted it for publication, having no idea how to go about it.

The Dead White Male syllabi of late-1950s and early-1960s English departments provided few accessible models to budding writers. The classics were classics for good reason. They were pinnacles of literary achievement in poetry, drama, the novel, the short story, and the essay. They were works to admire and study but impossible to emulate. Budding women writers, writers of color, were in an even more marginalized situation for a somewhat different reason. Their potential models were still on the edge of the canon, not in it, or not quite in it. But in perhaps every English department of the era there was an underground canon circulating among students, then consisting of Burroughs, Baldwin, Plath, Sexton, and other living writers. At Chicago, a few of us passed around Lawrence Durrell's *Alexandria Quartet*, which was being released in paperback one volume at a time. Melodramatic, flirting with taboo subjects, but with enough literary flair to be almost respectable, its many flaws suggested vague and exciting opportunities for young writers. Later, during grad school at Berkeley, the hot underground work was William Golding's *Lord of the Flies.*

After Berkeley, an old Chicago friend and budding writer, Ray Kingsley, helped me get a job as a technical writer with Space Technology Laboratories in Manhattan Beach. STL did systems engineering work for Atlas, Titan, and Minuteman missiles. I soon adopted Kingsley's plan of saving up $5,000, enough to live and write for five years in the cheaper corners of Europe, notably Greece. The dollar was king. The plan was much accelerated when I was given a choice of being either laid off or transferred to Norton Air Force Base in San Bernardino, for which I would be paid an additional $1,400, a year's savings.

My job was boring. It consisted of straightening out convoluted sentences and correcting neologisms, at least until I realized that the engineers were struggling to find words that described the rapidly evolving new computer technology. My editing work filled the morning at most. That done, I read the *New York Times* beginning to end. At the time Ian Fleming's James Bond thrillers were becoming popular. Out of curiosity I picked one up—and soon became an addict. The writing was bad, which is to say accessible, but the images and melodrama were compelling enough to keep me turning pages. Fleming led to Hammett, Chandler, and other detective writers. I was entranced by the narrative energy of these works and their patently manipulative techniques. Hey, I thought, I could do this.

In Redlands, in an old rented house, I wrote two short novels. The first, *A Deadly Device*, was a James Bond spoof about a Chinese plot to flood the American market with defective condoms, a strange harbinger of my fight fifty years later against a Chinese importer of cheap garlic. A second was a coming-of-age novel. I wrote both before and after dinner, under the influence of liberal doses of nicotine and port and sherry. I no longer have copies of either work. Both have turned to dust or ash somewhere.

In the spring of 1964, I made it to Lesbos. I rented a grand two-story stone house high up in the village of Molyvos, with a west-facing view of the mile-long arc of a shingle beach fringed by citrus and pistachio groves. To the east, behind the hill upon which the village perched, the Turkish coast lay four miles offshore. (In 2015, Syrian refugees crossed this narrow strip of the Mediterranean at the rate of three thousand a day). In Molyvos I joined a small colony of American, British, and French expats. In the middle of the summer, William Golding himself was to visit with his wife for a couple of weeks. A year later, I spent a night at their house outside of Salisbury. The medieval construction of Salisbury Cathedral was the subject of his latest best-selling novel, *The Spire*, which he jokingly referred to as "an erection in Salisbury."

Before Molyvos, I had never met a professional writer. By the end of the summer I finished two drafts of my first published novel, *GASCOYNE* (the protagonist-narrator refers to himself in all caps), a detective-story spoof. English classicist and writer Peter Green, then a full-time resident of the village, helped me find a publisher for the novel. Two years later, when I was living on Crete, Tony Curtis bought the movie rights. My new Australian wife, RoseMary, and I then bought the two acres that would become El Bosque Garlic Farm. Over the next forty years, seven more of my novels were published. The movie has never been made.

That's my experience with fiction. Our El Bosque Garlic Farm, though small, is not fictional. You can taste its crops, even its earth, as once used to be the habit of farmers. You can touch its fences and gates and tools and machines. You can even shake hands with its owners and workers.

But what the real El Bosque Garlic Farm can't do, as a complex fact, as a sprawling material presence, is walk into a McDonald's and order a cheeseburger.

Chapter 10

DRIPS

Back in 2014, when I filed the first Administrative Review Request of garlic importers with the Department of Commerce, I didn't have a sense that my little farm could be much of a threat to a large importer of Chinese garlic, and I didn't have much time to think about it. I had more down-to-earth problems to deal with.

Late winter is when the farm begins to waken. One of the first major tasks is to spread a mulch of wood chips over the garlic and shallot beds, the purpose of which is to retain moisture in the soil and help suppress weeds. The chips come from pruned branches and stalks of our own trees and shrubs, plus several dump-truck loads hauled to the farm by Robert Lopez, who specializes in clearing branches and trees around utility lines. Once we have a good pile of chips, I fill up three large harvest bins with the tractor front-end loader, three and four scoops each, and then slip the rear-mounted forklift under the bins and trundle them out into the field. I drive the tractor slowly up the beds, as guided by lines of emerging garlic shoots, while two workers spread the chips with pitchforks. It is the work of two days to mulch completely our acre-and-a-half patch of garlic and shallots.

Before the spring mulching, we scatter wood ashes from the fireplace over the beds by hand, plus measured amounts of Texas green sand, an ancient marine deposit rich in iron, phosphorus, potash, and trace minerals. We also spread humates, which are good for increasing the water retention capacity of the soil, and either soy meal pellets for nitrogen or, most recently, a mix of seaweed and bat guano. We're not a certified organic operation, but our practices reflect the requirements of certification. I'm a firm believer in feeding the soil, not the crops. Though the mulch requires nitrogen to break down, the process is slow. Since we have mulched with wood chips for about ten years now, early applications are breaking down and releasing back their nitrogen.

After mulching we lay down drip lines. As a long-time flood irrigator with water from the irrigation ditch above the house, I was a slow adapter to drip irrigation. I couldn't see myself crouching in the hot sun fiddling with little bits of plastic. Younger farmers had quickly taken up the practice when equipment and supplies became readily available through a couple of catalog operations. Fortunately, I was not farming during the record-breaking drought of 2002 and 2003, but the drought made me rethink my opposition to drip. A neighbor had installed a simple, easy-to-understand system. I decided to try it on our then-small garden patch. Drip tape delivers small quantities of water right next to seeds and seedlings via fine baffled slits eight inches apart. A first result was vastly improved germination. As I gradually returned to farming in the years following the drought, I expanded the drip system to cover all our row crops. This required thousands of feet of drip tape and a larger pump and filter.

By then the advantages were clear. Though you need to irrigate fairly constantly, you use relatively little water. Our acre and a half of garlic and shallots was first irrigated via a 1.5-inch-diameter plastic pipe down from the acequia, the flow boosted by a small irrigation pump and filtered by a swimming pool sand filter. This versus three 4-inch-diameter pipes needed for flood irrigation. Under a USDA Natural Resources Conservation Service grant, the system has since been upgraded to include a 3-inch underground main line and a 10,000-gallon storage tank.

With flood irrigation you need to be in constant attendance. Feeder ditches become clogged or break. The guiding of water down furrows next to crops can be time-consuming, exasperating, particularly if there are gopher holes or blockages from leaves and twigs.

With drip you can check the header lines and drip tape for leaks, and then go do something else. Pump turned off, with the system under gravity feed, you can go to town or let the drip run all night. With drip you can lay mulch, which with flood will create little dams in the furrows. With drip I stopped cultivating for weeds with the tractor, leaving the earth soft and mattress-like. With drip you get water right next to the plants without waiting forever for capillary action to move moisture close to the roots.

No system is perfect. There are problems with drip. Ravens love to poke holes in the tape. The tiny drip tape emitters, eight inches apart, can become clogged despite the sand filter, at times blocking stretches of various lengths, which are hard to detect when plants become large. The problem will be mitigated by the new 10,000-gallon storage tank, which will help settle sediment. Ideally drip tape should last two or three years, but then you're stuck with garbage bags filled with it, which can be recycled in major agricultural areas such as California's Central Valley, but not yet in Northern New Mexico. Fortunately, a fellow farmer occasionally calls on the farm and picks up the sacks. He uses the old drip tape to weave baskets and hammocks.

Drip also adds to material and equipment costs, though I have been able to raise my garlic prices to more or less keep pace. Perhaps educating my customers about the machinations of our Chinese opposition has helped. When we bring our first garlic bulbs, the Turban Chengdu, to market in May, along with bundles of small green garlic, we are often welcomed with exclamations of "At last!" and complaints about the poor quality of supermarket garlic available during the winter.

Chapter 11

WATER, WATER, WATER

A*cequia* is an Arabic word. In Northern New Mexico it refers both to an irrigation ditch and to the age-old administrative structure by which the ditch is governed, consisting of three *comisionados*, or commissioners, and a *mayordomo*, or ditch boss, all elected by the landowning members of the ditch, the *parciantes*. The word *acequia* harks back to the golden age of the Iberian Peninsula, when Christians, Jews, and Muslims lived in fitful harmony with each other and engaged in rich cultural exchanges, and when Arabic was the language of culture for everyone. The Moorish North African irrigation customs migrated to the future Spain and then crossed the Atlantic during the Conquest, finding a new home in Northern New Mexico, where a thousand acequias have operated under the flags of the Spanish crown, Mexico, and the United States. My acequia is small. Its three-mile-long channel serves some forty small properties.

The channel is fed by a makeshift diversion dam of rocks and brush across the Río Embudo about a mile east of the farm. Of the six acequias above it, three feature diversions of permanent concrete weirs. After spring flooding or summer flash floods, the temporary dams have to be rebuilt by hand and generally tightened up when the river drops in early summer, before the monsoon rains start—when they come at all.

The acequia is what fills the 10,000-gallon storage tank and supplies my 13,000 feet of drip tape with water, a nice conjunction of ancient and modern. The purpose of the tank is to give us a reserve during tight water times. For later use, it can be filled at night when others aren't irrigating or during my assigned hours when a rationing schedule is in effect.

Our household water now comes from a well toward the southern fence line, some seventy feet from the river, where it taps into the subterranean flow. Part of converting the Tower into a guest house consisted of installing a new septic tank to serve both house and Tower. Rather than continue to drain effluent via a drainage field into the ground, I decided to build a "constructed wetland." This consists of a thick gravel bed planted with cattails and bulrushes on top of an impermeable liner. Effluent pumped up from the septic tank flows through the 10-by-30-foot gravel bed and drains into a 10-by-10-foot pond, two feet deep, stocked with gambusia or mosquito fish and water hyacinth during the summer months. I regularly flush out the pond with acequia water during the summer to keep ammonia levels down, pumping the pond water onto the lawn and trees, not row crops. During the winter when it's too cold to operate the sump pump, the pond drains back into the old drainage field. As effluent, it is much cleaner than what comes directly out of the septic tank.

The wetland also enables us to use household water twice, important during drought years, as a way to keep our small lawns and shrubs and trees and flowers alive.

Chapter 12

WEEDS, WEEDS, WEEDS

Mulch and drip don't eliminate weeds, though mulch cuts down their number. After a short, warm winter, I can be out in the field with a scuffle hoe catching the first emerging weeds as early as March. By now I know them almost as well as my crops. Grasses come first. They favor cooler weather, and it's good to catch them before their root crowns reach deep into the soil. Their narrow leaves resemble garlic: they're easy to miss. The scuffle hoe features a thin double-edge U-shaped blade that cuts by either pushing or pulling, but often the blade fails to snag clumps of grass, which then must be pulled by hand.

Mallow, a sprawling plant with round leaves, is easy to pop out of the ground while young. Fully mature, it roots so deep that often you can't pull it out of the soil by hand unless the soil is very damp. When young and tender, lambsquarters or quelites, as they are locally called, are regarded as a spinach-like delicacy by my Hispanic neighbors. Other early weeds are pigweed and wild amaranth, which, like lambsquarters, can grow two and three feet tall if allowed to do so, whereupon only a string trimmer can deal with them.

Purslane, or verdolaga, with fleshy light green leaves and pink stems and tiny yellow flowers, begins to sprout in the warm days before garlic harvest; it, too, is edible. Late arrivals after garlic harvest are goat-heads, or cabritos, usually appearing on the edges of driveways and footpaths, with inconspicuous fringes of tiny leaves. Fortunately, their small yellow flowers give them away. Their three-pronged spines are painful, can puncture a bike tire. After digging up the sprawling plants, I drop them into the trash barrel.

Ideally you want to weed even before most weeds emerge, by disturbing the soil so as to disrupt the growth of sprouting seeds. And ideally also you want to eliminate weeds before they go to seed, producing yet more generations. But it's a hopeless task. Weed seeds can lie dormant in the soil for years. Some are windborne, others are brought into the field by irrigation water, though in theory the drip filter should catch them. I've had to train interns to thoroughly disturb the soil, not just hoe out the bigger weeds, leaving the small ones for "later" when they will be big. . . .

Weeding can be meditative work, at least until your back complains. Thoroughness can be gratifying but involves going back over a row of a bed in the opposite direction in order to detect seedlings hiding behind garlic plants. And it's gratifying to catch them all, even if it requires bending over or kneeling to pull out a weed growing right next to a garlic plant so close you can't get at it with a hoe.

When the garlic plants are fully grown, much weeding will be on hands and knees. I add a hand weeder with a sharp triangular blade for working close to plants and dealing with clumps of emerging seedlings with a sweep or two. I also move a bucket along with me to collect grasses and weeds for the chickens. What they won't eat, they will scratch eventually into compost.

We weed up to the last minute before harvest. Failing to do so leaves too many large weeds that can foul the undercutters, forcing me to stop the tractor and climb down and disentangle them from the shafts of the undercutting disks. In the last week before harvest, we'll concentrate on only the large weeds, leaving the small ones to be dealt with by the undercutting process itself, which leaves a forty-inch swath of disturbed soil.

Chapter 13

THE GARLIC FARMER AS HERO (TO SOME)

In the early spring of 2015, during one of my first weeding sessions, I got a call on my cell phone from Ted Hume. I had a sense that something was up from some of his ambiguous remarks of late. Ted speaks with a deep, warm voice. At that point I was still deciphering the arcane terms of anti-dumping law. Ted had decades of experience with this business, and sometimes his acronym-sprinkled remarks (AR 20, APO, POR, PRC) came across as a kind of shorthand. His manner of speaking was also lightly ironic, posing the question of whether I should take his remarks at face value or parse the meaning of his subtle spin. At least by now I understood that I was involved in a complex situation of still-unknown dimensions.

All I knew clearly was that as an American producer of garlic, a "like product," in US Department of Commerce terms,

I had the right to request a review of the duty paid (or not paid) by any and all importers of "fresh garlic." The issue was that Harmoni was spending large amounts of money contesting my "standing" as a commercial garlic producer. In my filings with Commerce, I supplied tax returns, employment records, planting and harvest records, loan and insurance documents, all to establish that I was in the business of growing and selling garlic. Yet they maintained I was only a hobby farmer, and that my operation was a "fiction."

During one of Ted and Renate's early-2015 visits to Taos, where they were thinking of relocating from Ojai, the three of us had made the rounds of the New Mexico congressional delegation offices in Santa Fe. We argued that cheap imported garlic had a negative effect on small New Mexico producers. We urged the members of the congressional delegation to write letters to the Department of Commerce in support of our request to review garlic importers. Staffers were very receptive to our concerns. During our time with Nicholas Maestas, head of Congressman Ben Ray Luján's Santa Fe office, Ted startled me by saying, "Mr. Crawford is quite courageous in this effort, as there are certain risks involved."

As soon as we left the office I asked Ted about these "risks."

"Oh," he said, "that Harmoni might offer you a couple million for the farm."

After a moment, I responded, "Well, I suppose I can live with that kind of risk."

In a phone conversation a few weeks later, when Ted hinted that something was about to happen, I said, "Look, I want to live out my life on my farm," which was to say it wasn't for sale even for an outrageous price. "And as a writer, I won't be silenced."

When he called during weeding season, a few weeks later, I was standing in the garlic patch, hoe in one hand, cell phone pressed to my ear with the other. Ted said, "There have been some developments. We have to withdraw."

"Withdraw?"

"Withdraw our request for administrative review of Harmoni. The manager of a Chinese client of mine is being threatened with prison. Harmoni is big enough to have government officials in its pockets."

I had no idea of what the process of withdrawal was about. I eventually learned that a petitioner—a domestic producer of the imported product in question—can withdraw a request for review with the Department of Commerce within ninety days of the beginning of the annual review period. This in fact is what Harmoni and Christopher Ranch had done every year for the past ten: first, like me, Harmoni (US) and Christopher Ranch filed a request for review of all garlic importers, and then toward the end of the ninety-day period, they withdrew Harmoni (China) from the list of companies to be reviewed. Everything then stayed the same. Or would have stayed the same if I also withdrew my request for review. Harmoni (China) would get another year of paying no anti-dumping duty. This practice of Harmoni and its California allies first asking for a review of Harmoni's duty rate and then withdrawing from the review process is, we believe, a legal gaming of the system, which we also believe could not have been the original intention of the law. The ostensible purpose of the anti-dumping law is to assess duties on imported products being dumped below cost into the US market. It makes no sense to legally enable the American collaborators of an importer to cherry-pick its Chinese importing partner out of the review process, in this case retaining its zero duty rate for ten years running. But this gaming of the system is precisely what Commerce has sanctioned.

"Well, I guess we should withdraw," I said, still not fully understanding the administrative mechanics of the situation.

A week later a check arrived from Ted for $50,000. This was a surprise. Other than the offhand remark about Harmoni possibly offering to buy me out, there had been no mention of money. Nor at the time did I have any idea of the huge financial stakes involved in the Chinese garlic importing business. I later learned that if Commerce ever reviewed its garlic imports and imposed the highest rate of duty, Harmoni could be subject to over $200 million in tariffs per year.

For a long time, I had no idea who the money was from, other than Ted, or exactly what it was for. For a while I even wondered whether it might be from Harmoni, trying to buy me off—except the money came with no conditions. Whatever—as a farmer I was chronically in debt and was happy to be able to pay off my credit cards, put a down payment on a new Kioti tractor, and expand my photovoltaic system. I was also happy to share the news of the bounty with my friends. Ted later said the money came from his personal account, as a gift in appreciation of what I had taken on. The IRS also considered it a gift, for which Ted paid the tax.

For some time, Ted had been fond of saying with his usual ironic twist that I was considered a "hero" in China. This seemed hyperbolic. Hero to whom? To how many people? And for what?

"For taking on Harmoni," he said.

Later that spring, Ted proposed that we travel to China in July, after garlic harvest, to meet some of his friends' former clients. We would fly to Beijing, spend the first night in the airport

Hilton, then take a high-speed train to Shijiazhuang, capital of Hebei Province, where Ted's old friend Ruopeng Wang lived and maintained an office. Decades before, Ruopeng had started out as Ted's translator during "verifications" of export documents for Commerce, becoming a close friend, to the point that Ted and Renate regard him almost as a son. Eventually he created his own business as an export consultant. The plan was that once we reached Shijiazhuang, Ruopeng would drive us to the huge garlic growing areas of Shandong Province. We would end up in Qingdao, on the coast, then return by train to Beijing, a ten-day trip in all.

Not counting a 1967 visit to Hong Kong and Macao, it would be my first trip to China.

Chapter 14

THE USES OF NONFICTION BOOKS

The proposed trip to China was another indication that my involvement in the Chinese export situation was leading me in unexpected directions. In preparing for the trip, I read a couple of books on China. A Mandarin phrase booklet? A quick thumb-through convinced me of the near impossibility of learning more than "thank you" and "hello," though I slipped it into my small carry-on anyway. I also packed a dozen books of my own, mainly *A River in Winter: New and Selected Essays*, still in hardcover, as gifts to my future hosts.

Probably no one would read the books, or be able to read them. I doubted the volumes would even lead to questions of how I became a writer of nonfiction after starting out as a novelist. Or how a young novelist became a farmer, and how that led me away for a time from fiction into the personal essay. The years 1968 and 1969 resonate with Americans of a certain age in ways that might seem trivial to Chinese people who at that time were enduring the brutal excesses of the Cultural Revolution, but the events of those years set RoseMary and me off in a new direction, to escape a Bay Area wracked with polarization over the war in Vietnam. A persistent gas leak released us from the lease agreement of our apartment above a Japanese restaurant, at the corner of Geary and Thirty-Sixth Street in San Francisco's foggy and monotonous Sunset District. In the fall of 1969, we packed ourselves and one-year-old son Adam into our VW camper van and headed out for Northern New Mexico to rejoin new friends who had recently made the move. The simple life beckoned—though not exactly "tune in, turn on, and drop out." A few experiences with pot had shown the drug was not for me. My life was a struggle to become articulate, and pot dropped me into a distressingly wordless void. But as a new husband and father I was becoming aware of my too-great dependency on what seemed to me to be an increasingly unstable industrialized world. I needed to learn how to do things myself. Garden. Build. Fix things.

Within six months of moving into a rented adobe in the Embudo Valley 50 miles north of Santa Fe, RoseMary and I started growing our own garden, my first-ever experience in regular physical labor. Within a year, with the last of the movie money, we bought our first two acres. The following summer we made the adobe bricks for the first two rooms of our house, into which we moved in December of 1971. The experience of building and gardening, soon to become farming, succeeded in smoothing over the rough areas of our still-young marriage: we bonded over our strenuous labors. For some of our young hippie friends, such efforts had quite the opposite effect.

Designing and building our simple adobe house was the most exciting thing I had ever done. Its challenges and gratifications exceeded whatever I was able to imagine in my fiction writing. Eventually I tried my hand at a series of essays about living in the somewhat foreign country that was Northern New Mexico at the time. Yet there was a big hole in my knowledge: Los Alamos Scientific Laboratory, 45 miles to the southwest, birthplace of the atomic bomb.

In our first years in New Mexico, I had driven up to the famous Lab salvage yard a number of times with friends, loading up the truck with inexpensive odds and ends—a 2-foot wide and 6-foot long redwood plank, which I still have, desk chairs, shelving, storage boxes, and a long-carriage Remington office typewriter. But I was oblivious to everything else. Many of my neighbors, both Hispanic and Anglo, worked at the Lab, but their work was not a subject of conversation. I had no clue what really went on up there. Up there because Los Alamos was a thousand feet higher than the Rio Grande Valley and its northern tributaries. Up there because the town housed one of the highest concentrations of PhDs in the country and because the per capita income of Los Alamos County was the highest in the state and indeed much of the country. A brief tenure in the mid-1970s as a writing subcontractor for a solar handbook, work I mostly did at home, got me no closer to the main mission of the Lab. And the offices of the solar project were in the neighboring suburb of White Rock, not within the boundaries of the twenty-three-square-mile Lab.

On the morning I finished writing a Los Alamos chapter of an ill-fated essay collection, "Rattlesnake Pueblo," I went for a late-morning walk along the river to ponder what I had written and what might come next. I wasn't pleased with the morning's work. At best I had succeeded in skirting my own ignorance of

Los Alamos. I returned home to find a strange car in the drive. Its middle-aged occupant was too frightened of our little terrier to get out. The car bore US Government plates.

From the driver's window, the man explained that he was here to ask questions about a woman who had worked for us on the farm. She had applied to become a security guard at the Lab, for which she needed a security clearance. He introduced himself as a federal security investigator.

Shooing away Pooch, I reluctantly invited the investigator into the living room. From our sagging couch he asked me a series of questions about our friend. Alcohol or drug use? Marital discord? Subversive political views? Anything else the Lab should know about her or her family?

I answered as briefly as possible, portraying her as the saint she was.

After he left, I was angry at myself for collaborating in a "security" system that relied on friends and neighbors and employers informing on each other. The visit had also driven home doubts about how little I knew about Los Alamos Scientific Laboratory. (Some years later "Scientific" was replaced by "National.") It was time to end my ignorance.

In the first months of 1983, I began spending a day a week poking around Los Alamos. The Lab roads amid the far-flung facilities across a canyon from the town were still public, and the main collection of the Lab library was easily accessible. I had a couple of acquaintances who worked for the Lab, and another who lived up there. I often had lunch in the Lab cafeteria. Back home after my explorations, I would wait a day or two before writing out an account, as sparked by some odd detail that stuck in my memory. Such as guys playing volleyball in front of the

Plutonium Facility. Or why a small town had so many churches, some thirty in all. Or the old entrance guard tower, a relic of the years 1943–1947, when Los Alamos was a closed secret city. Or the curious business of naming atomic test blasts after flowers and plants and animals, from a list I found in the reference section of the library stacks but which later was removed.

In effect, I was teaching myself how to write the kind of nonfiction in which the writer-observer plays a part in what he or she is observing. I amassed a thick typescript of such observations over the course of six months. Finally, I realized I didn't have the science to delve deeply. Nor did I wish to devote more of myself to the intricacies of the Los Alamos death machine. I abandoned the project. Or not quite. It all came back ten years later in condensed form, when the US entered the first Gulf War over the Iraqi invasion of Kuwait. In a rage over American involvement in a fight between a totalitarian regime and repressive sheikdom, I pounded out a brief chapter called "The Atomic Bomb Ring," which would later appear in my book *A Garlic Testament.*

Mayordomo: Chronicle of an Acequia in Northern New Mexico was the first book-length fruit of my Los Alamos training. It was sparked by a remark of an old family friend, Alister Brass, an MD whose professional career had taken him into editing medical and scientific journals. Long before, RoseMary had worked with his journalist father in Australia. On his March 1986 visit to the farm, I led Alister up the ditch bank fifteen feet above the backyard to show him the day's work. He was a trim fellow in his late forties with sandy blond hair, with an accent as much English as Australian. By then I had been mayordomo, or ditch boss, of the Acequia del Bosque for a number of years. Up on the ditch we looked down at the newly cleaned 4-foot-wide dirt channel.

It wandered through the cottonwoods and volunteer cherry trees and was overhung by junipers and New Mexico olives, at the base of a 100-foot rocky slope. The irrigation ditch marked the border between a riparian ecosystem to the south and a piñon-juniper woodland life zone to the north. The incongruous presence of this winding dirt channel, still empty of water, suggested a creation by the environmental artist Andy Goldsworthy.

"Stanley, you should write about this," Alister said as we stood on the ditch bank. He was an avid and perceptive reader. He always brought a stack of new books for us to read, most recently the magnificent *Raj Quartet* by Paul Scott.

After a while I replied, "I certainly know enough." By then I had worked on the ditch for almost twenty years, first as an elected commissioner, then as mayordomo. "But I don't know how to write about it."

Alister left the next morning, the last day of the three-day ditch cleaning, in which a crew of twenty-some guys sliced vegetation off the sides of the ditch banks with sharpened shovels and dug out patches of the channel that were silted up. His words stuck in my mind. I began watching and listening to the ditch crew with new eyes and ears. And I stepped back and observed my place in this ancient ritual, a newcomer Anglo who had been involved with the acequia and its Hispanic members since the mid-1970s.

That day and the rest of the spring and summer, I recorded my observations in the back of a notebook, in pencil, as if to tell myself I wasn't really writing about my adopted community. Nor did I tell anyone what I was writing, uncertain as I was about the project. Not even Alister.

One of my great regrets is that he never knew about the manuscript. Toward the end of the year, he abruptly quit his

San Francisco job and returned to Sydney, where he entered a hospice. In our last phone call, he said he didn't want us to fly over to Australia to be with him. He died of AIDS less than a year after our March visit. With sad irony, he and another doctor had just published a book about AIDS.

The ditch book, as I call it, was published two years after Alister's death. I dedicated it to him and to my parents.

Chapter 15

ABOUT NUTS, ABOUT BOLTS

We scheduled the China trip for July. This gave me enough time upon return to harvest garlic and shallots and to oversee the planting of the field in winter squash and greens. And, before departure, to deal with a long list of spring tasks, many of which were left over from the autumn. An ideal list would include brushing linseed oil on the wooden handles of shovels and hoes, changing oil in the small gas engines that power various mowers and the two-wheel tractor, repairing and restoring and repainting the 1940s Palmer apple boxes we use to store and transport produce, replacing worn tines in the tractor-powered rototiller, fixing a few minor problems with the pickup, and cleaning out the sheds in preparation for storing garlic and shallots.

Of those, replacing the rototiller tines is the largest task. The tines wear down every few years to thin L-shaped blades prone to breaking off. New blades are a quarter-inch thick, three inches wide, and about a foot long, with a 90-degree bend toward the middle, the beveled cutting edge spotted with beads of hardened steel. There are sixteen tines on my tiller, each held on with two sets of nuts and bolts. To change the tines, you attach the tiller to the tractor, raise it, and secure the 200-pound implement in the upper position. Then, on hands and knees, you unbolt each of the tines. Usually about half need to be replaced at the same time.

I find this strenuous but gratifying work. The tines are the product of industrial processes I more or less understand. In my youth on family road trips around the country, we visited the Maytag washing-machine factory in Newton, Iowa, and the vast River Rouge Ford Motor complex in Detroit, where silvery body panels were transported by aerial conveyor belts to assembly stations. In drives through the western states, I awaited those moments when we would pass steel mills and mines and smelters. Up until the 1960s, industrial and consumer products were put together with rivets and screws and nuts and bolts. An inquisitive and somewhat handy boy could take apart almost anything. My father, a high school teacher of wood, metal, and electric shop, was the household fixer and builder of things. Up until the 1990s, backyard mechanics could still service and tune up their vehicles.

Such is still true of most basic farm equipment, whose mechanisms consist of blades, gears, chains, levers, and wheels. When I show off my Polish automatic four-row garlic planter, most people understand very quickly how it works. There is nothing mysterious about rototillers, cultivators, mowers, scraper blades, and the like. Besides nostalgia, this is the appeal of old and antique collector cars and tractors. The advent of YouTube step-by-step videos on virtually every possible topic has made servicing old equipment even easier. We may still be blind and clumsy regarding the microbial tumult that characterizes healthy soil at the microscopic level, but we're very good with machines that manipulate the soil.

The mechanical is what I would call a comforting factor in farming: you are dealing with processes and equipment that are still understandable, at least on the gross level of human eye and hand and foot coordination. The basic operations of my farm will not be affected if the internet goes down—or simply when, as some argue. Political and financial upheavals won't prevent me from tilling or harvesting or planting my fields. The same can be said of those artists and artisans and craftspeople who still work with their hands and basic hand tools, besides many in the building professions.

Farming, like many artisanal professions, has feet in two worlds: in prehistory, before written records, when plants and animals were domesticated; and in the rapidly evolving digital world. Most farmers, even the Amish, now use the internet, and while large corporate farms employ implements guided by GPS and laser systems, the time may come when even the smallest farm will rely on these technologies.

Chapter 16

INNER SPACE AND OUTER SPACE

I find it initially disorienting to enter airports and jet airliners: those environments devoid of the presence of nature except at a great distance through double-glass windows. Once inside these spaces you are surrounded by concrete and steel and plastic in its many configurations, remote from nature's myriad irregularities and from the feel and smell of the earth and plants and animals. My personal space in house and farm is some two acres, plus the half-mile walk I make twice daily along the river, against a backdrop of rushing water, with the two heelers, Tippie and Bucky, darting in and out of the bushes to either side of the lane—four-footed extensions of my own senses.

There is something deeply shocking to the psyche to be confined to a space a foot and a half wide, two and a half feet fore to aft, and five feet floor to overhead luggage bins—though the fore-to-aft space is larger in bulkhead seats. Almost as shocking is how readily we submit to such contractions of personal space, in the interests of overcoming vast distances in short periods of time. But also how, on those few occasions when I have flown frequently, I feel relieved of all responsibilities while so confined, and even pleased to be so.

The flight to Beijing from San Francisco was about fourteen hours. I had a bulkhead seat next to the galley, with two empty seats next to mine. Ted was about fifteen rows ahead, on the opposite side of the plane. Departing later afternoon, we flew into an early evening and a longer-than-normal night, arriving the morning of the next calendar day in Beijing. After a night at the airport Hilton, Ted and I took a taxi to the main Beijing train station, a sprawling complex of forbidding proportions. In the past the hotel would have obtained train tickets for its customers, but now we had to find the right booth. After several tries inside and outside, Ted leading the way up and down stairs, we found the correct line and bought our tickets. The vast waiting hall was filled with thousands of people milling around. At seventy-seven and six foot three, I estimated that I was the oldest and tallest person there.

During the hour-long trip, the eight-car high-speed train to Shijiazhuang attained 293 km/h, or about 175 mph. In the course of our week in China we would pass perhaps fifty of the trains on this and the Qingdao-Beijing lines, a small portion of the high-speed system in China. On the one-hour trip, once we left the high-rise suburbs, we passed endless fields of corn. The plants were four feet high, as perhaps they would be back in Iowa at this time in July. The air was murky, a mix of mist and smog, but not oppressively so.

Ruopeng Wang greeted us at the enormous modern Shijiazhuang station. He was a short, smiley fellow in his fifties, with a thick brush of black hair and sideburns trimmed short.

His English was reasonably good—far better than the halting efforts of the hotel and dining room staff we had chatted with.

We spent two nights in central Shijiazhuang at a high-rise hotel near Ruopeng's office. Visibility was perhaps a half mile, but few cyclists wore masks. Shijiazhuang is the capital of Hebei Province, which I later learned is the source of much of the industrial pollution that regularly engulfs Beijing. Ruopeng regaled us with meals in private rooms of a number of restaurants, plus a lunch in his office dining room along with the staff of his export consulting firm, including his attorney, Lucy. The meals were all centered around huge lazy Susans piled with a staggering variety of dishes. After our first lunch together, we went upstairs to a conference room, where we discussed the business of exporting Chinese garlic to the US. During the rest of our week together, I found it puzzling that there was no further discussion of garlic exporting matters.

Ruopeng had drafted an old restaurateur friend, Dong, who spoke no English, to accompany us all on a several-day drive to Qingdao on the coast, with two overnight stops in Qufu, home of the extensive Confucius compound, and Jining City, to visit a garlic-processing plant. Garlic harvest in Shandong Province had been completed the month before, and the land had been replanted mainly with corn. As of 2014, China was by far the largest producer of garlic, at 19,233,800 metric tons. By comparison, the US production was less than one percent of that, only 175,445 metric tons.

With Dong at the wheel of Ruopeng's Audi station wagon, we traveled by well-designed superhighways between cities. Video cameras were everywhere, successfully discouraging speeding and running red lights. Other than countless variations of electric bikes and scooters and tiny, boxy three- and four-wheeled two-passenger taxis painted bright red, and of course the food, there was little that suggested the kind of radical foreignness that Western travelers would have experienced a generation or two before. Architecture, dress, cars could have been dropped anywhere, minus signs in Chinese characters.

The garlic-processing plant is out in the country from Jining City and is owned by a former farmer, a Mr. Wang. It's a huge complex housing more garlic than I could grow in several lifetimes. Garlic is peeled, cleaned, and sorted in a spotless room, entry into which requires that we don masks, booties, caps, and overalls.

After a night in a monumental Jining luxury hotel slathered in marble and whose design suggested that it was inspired by the Vatican, we drove on to Qingdao, the former German concession city on the Yellow Sea, where we were hosted by exporter Jack Bai. Qingdao is a very Western city, with clear skies and leafy boulevards plied mainly by cars and buses, minus the swarms of bicycles and scooters common in the other cities we had visited. On the long attractive sea walk that winds in and out of rocky coves, young and old walked and jogged, many with dogs on leashes—which we had also not seen elsewhere.

After several meals with Jack Bai, whose English was minimal, and two nights in the grand Sea View Garden Hotel overlooking the Yellow Sea, where one might easily imagine they were in Cannes or Nice, Jack drove us to the train and we headed back to Beijing.

I found the trip fascinating but felt somewhat confused by what it was all about. During our last dinner, at a Japanese restaurant in a mall in downtown Qingdao, Ted tried to turn the conversation to business, but without success.

On reflection, it seemed to me that our Chinese colleagues were sizing me up. Ted had known Ruopeng for decades and has been Jack's legal representative on export-import matters. At one point during our travels, Ted pulled me aside and reported that Ruopeng and Jack thought I was a nice fellow, an intellectual in their eyes, but that by implication I wasn't up to the battle with Harmoni. Earlier he had told me that Ruopeng was confused about why I was asking the owner of the garlic-processing plant about planting and harvesting dates, temperature highs and lows, farm equipment, fertilizers, and the like. It later occurred to me that these businessmen were interested only in container loads of garlic and shipping costs and the like, not in the processes by which the crop was grown. Those were peasant matters beneath consideration.

I thought at the time the non-fiction volumes I took to my new Chinese acquaintances were appropriate gifts. But I was wrong. In China, farmers are farmers, not writers, not university grads, not speakers of several languages. My gifts were confusing, I later realized. Unbeknownst to me at the time, they could have had the effect of derailing our whole effort against Harmoni. Our Chinese acquaintances were probably wondering whether they had allied themselves with the right guy to be their "hero."

Back at the San Francisco airport, while awaiting our Denver connection, I suggested to Ted, "This thing is all over, isn't it?"

"Oh no," he replied. "We're just beginning." Though he didn't make it clear what this particular beginning was.

This was July 2015. In the end, we would re-enter the fray by filing a request for review of all garlic importers, including Harmoni, in late November, the second annual review we were to participate in.

Chapter 17

BACK ON THE FARM

On my return home, recollections of the Chinese automated garlic-processing plant made our operation look minuscule, yet I was comforted by the familiar embrace of the mountain landscape of Northern New Mexico, the Río Embudo and the Río Grande, the flawlessly clear skies. Based on what we saw from the high-speed trains and from Ruopeng's Audi, hundreds of miles of corn and other crops, the productivity of Hebei and Shandong Provinces was astounding, though the lack of field workers suggested that the fields were mainly tended, as it were, by herbicides and pesticides. In the high-speed train, we would have passed the equivalent of the agricultural areas of all of Northern New Mexico in less than a few minutes.

The visit did not inspire me to become an industrial grower, even had I had the means. I was content with my small operation. As far as processing went, my only form of automation was a brusher-cleaner, a machine with a dozen motorized tubular brushes that tease dirt and loose skin and some roots from garlic and shallot bulbs. It's far faster than cleaning bulbs with thumbnails and fingernails, though it doesn't do as good a job. Yet bits of darkened skin reassure the buyer that our garlic comes from our farm. Commercial garlic is often so white as to suggest that it has been bleached.

Following the China trip, I resumed the summertime routines. These begin with unracking the garlic in the sheds, a bushel-basket-full at a time. With the Russian Red and Siberian, seed tops are snipped off. The bulbs are sorted into two lots. The better-looking bulbs of a good size will become half-pound decorative bunches wrapped with colored twine. Onto a stalk will be tied a half-dollar-sized pottery medallion made by a neighbor ceramicist Al Tyrrell, along with a farm tag. RoseMary used to make the decorative bunches with the help of two neighbor women, but we now farm out the work. The largest bulbs will be saved for replanting. The rest will have their stalks snipped off and will be sent through the brusher-cleaner once or twice, preparing them for the farmers' market. The softneck varieties, Turban Chengdu and Bosque Early, will be similarly trimmed and selected for market and replanting, as will shallots.

This is slow, meditative work, not unpleasant even in the heat of summer, in the shade of the maple that stands in the center of the driveway turnaround or the weeping willow that towers over the hoophouse. Once cleaned, market garlic goes into the Palmer boxes painted blue and turquoise and into the back of the pickup or van. At the market, these are unloaded into smaller display boxes and baskets.

In the past, we'd collect garlic seed tops in bushel baskets. For a couple of seasons, a Colorado grower paid me a modest price for them. But most would end up in the compost pile. Finally, we decided to see what would happen if we brought a basket to market and gave away handfuls as samples. We suggested that customers could use the grain-sized bulbils in soups, salads, on pizza, and guacamole, without needing to strip off the very fine skins, or even swallowed whole for medicinal purposes. The basket of rose-colored bulbils attracted attention. "What are these? I've never seen these before." Brave souls popped them into their mouths. "Tastes like garlic! Wow, they're strong!"

Once you engage people, they are more likely to buy something, even if not right then. And everyone likes a free sample. Eventually we took to bagging up pre-weighed bulbils in fine mesh bags. We had started doing this earlier with garlic and shallots, in order to spare hurried customers the time-consuming ritual of weighing and pricing handfuls of bulbs. To each bag we added a "wooden nickel" printed with the farm name and logo. Marketing, marketing. But such packaging adds waste. In the future we'll offer customers who return the reusable bags and the wooden tokens a dollar off on their next purchase.

Chapter 18

A BRIEF HISTORY OF GARLIC PLANTING

We first planted our garlic back in the mid-1970s by hand, carving a small furrow in the recently tilled earth with an arrow-shaped hoe.

I have never been one to worry about planting the cloves "right side up" with the root germ at the bottom, but there are growers who still seem obsessed with the issue and spend much time and effort hand inserting the clove germ-down. When the soil freezes and thaws in the winter, the action will serve to right the cloves as roots work downward, as I have written elsewhere. In mild climates, the dampening and drying of soil will have the same effect. Those cloves that fail to right themselves will still send up seed stalks though with bends in them, some as much as 180 degrees.

As we grew more and more garlic in the late 1970s and early 1980s, I took to opening planting furrows with the tractor, dropping the cloves into place by hand, and then moving the furrowing shovels over a few inches to cover the cloves. This was a rough system, but it worked well enough back in the flood-irrigation days.

In 1986, when I traded in the two-cylinder Kubota for a larger 30-horsepower four-wheel-drive model with a front-end loader, I also bought a one-row Holland transplanter, which we used for two types of planting. For seedlings, two workers sat on the machine and fed the plants into rubber grippers on a wheel. When the grippers revolved to a point just above the earth, they opened and dropped the plants root downward into a furrow the machine had opened. Canted packing wheels then closed the earth around the plants. Ideally, at least. In practice, many plants were dropped at odd angles, some even buried, requiring us to go back over each row by hand. However, the machine was still better than planting everything by hand.

For garlic and other bulbs, including potatoes, you had to spend the better part of the morning unbolting the dozen grippers and then bolting on small scoops. Garlic was easier than seedlings, but for the two workers who leaned into each other over the revolving scoops, it was still a chore. The long, gangly machine drifted side to side. As a result, you could only plant two rows per bed, which in fact was all you could irrigate with the flood method.

About ten years later, I replaced the Holland with a carousel-type planter made by Lannan of Finland. This required a single worker to drop cloves one at a time into a revolving eight-tube carousel with flaps at the bottom end of each tube. At the right point, each flap opened and dropped a clove into the planting furrow, which was closed by the canted packing wheels, same as the Holland. The carousel was turned by a long quarter-inch-thick rubber band attached to the packing wheels. Surprisingly, it endured ten years of New Mexico sunlight before having to be replaced.

When we converted to drip irrigation, we were able to use the Lannan to plant four rows per bed, not just two. We first planted the two inner rows, after which we slid the machine six inches to one side on the toolbar to plant the outer two rows. The process was somewhat nerve-wracking, as often it was not readily clear which beds had been planted with two or four rows, particularly after a rain and when the dogs had run over the field. A heavy fall of autumn cottonwood leaves could also obscure the tracks.

Almost another ten years elapsed before I found online a four-row automatic garlic planter from Poland. It consisted of a large hopper through which four rows of shallow cups rise. The cups are slightly dished washers welded to a chain, and the four chains are powered upward by a set of rubber tires. Ideally each cup picks up a single clove and drops it down a tube, into a furrow opened by a small furrowing shovel. Four packing wheels push the cloves deeper into the earth, and a round bar covers the cloves with soil, leaving a smooth bed behind. But "automatic" is something of a misnomer. The machine is equipped with two seats from which workers fill the occasional or even frequent blank cup or else knock off cloves from cups that end up with two or more, often the case with softneck garlic.

The four furrows are twelve inches apart, which presents no problem in soil free of stalks or an abundance of fresh organic matter. If the soil is filled with “trash,” then we have to stop and raise the machine and rake out little dams of stalks and leaves. But even so, I estimate that the machine is eight times faster than the Lannan. And it’s generally easier on the workers riding it.

Chapter 19

NAKED AFTERNOONS

I have no reason to complain about summers devoted to farming, even though I can often get away for only a night or two. My schoolteacher parents lived for their summer vacations, during which they explored the western national parks after World War II in a succession of camping trailers and small travel trailers, their last being an Airstream. During my four writing years on Lesbos and Crete, daily afternoon walks or drives to the beach were the norm from May through October. I took my vacations when young, worked when older. Though for several years, RoseMary and I were able to get away to San Carlos and Mazatlán and San Blas and San Miguel de Allende for a month at a time in winter.

Fortunately, the small Río Embudo runs along the southern border of our property, across a dead-end lane at the bottom of our largest field. Almost every afternoon from late June until early October, I walk down the drive and down the lane a short distance to a tall fringe of willows and thread my way through them to a sunny, grassy spot next to the water. You wouldn't call this a swimming hole so much as a dipping spot. During much of the summer, the 20-foot wide watercourse is usually at most eight inches deep except in places where I roll boulders in rough lines to sluice the flow into a deeper channel, a bathtub of sorts. Up until three and four in the afternoon, the summer sun shines directly here above thick corridors of cottonwoods and willows lining the rocky watercourse. Flycatchers twitch on overhanging branches. In late June the water is still cold but no longer icy. Silty flows after flash floods can be almost warm.

The river is subject to extremes, though not every year. During the severe drought of 2003 and 2004, the flow was reduced to a trickle between small stagnant pools, orange with algae. The drought of 2018 was even worse, drying up the river completely for most of the summer. Several years ago, record August flows were so high you couldn't step into the water without risk of being swept away. More recent summers, except for 2018, have been closer to what I think of as normal. As the flows diminish in late July and August, monsoon rains and afternoon downpours bring the river back up.

I don't spend a lot of time down there each afternoon, perhaps half an hour at most, just enough to slip out of shorts and shirt and immerse myself in the cold flow, sit for a while in the water, then dry off on the grass, and take a last quick dip before slipping clothes back on. Before or after in my Teva sandals I inspect sections of the river for flat stones, which I carry up to a pile next to the gate for paving future pathways.

The spot is private enough that I rarely put on the swimming trunks I bring along just in case. Judging from the footprints in the sandy patches, I'm the only one who ever visits the river here, along with my dogs. Very rarely do I see the boot print of a presumed fisherman. Pleasant as this may be, it's also worrisome that so few people seek out the restorative presence of the river, the quiet, the gentle musical sounds of the water threading its way over and around rocks, the calls of songbirds hidden away in the upper branches. In these times there is something special about being away from the complex apparatus of contemporary life, its insistent digital presences, and being in the company of just nature, just sky, just water, trees, grass, ants, water skaters, newly-hatched fish fry, and the rubble of ancient geological eras in the form of countless different types of rocks tumbled together over the millennia by the action of earthquake, flood, the freezing of water into snow and ice and its liberating thawing. Sitting naked on the grass amid this presence, with civilization in the form of rumpled clothing parked aside, I know I am a small thing, a small creature, a mere speck amid the universe, strangely endowed with a mind that can project itself deep into the starry universe, reach far back into human history, back toward the origin of the universe itself, and also gaze far into an imagined future.

Clothes back on, watch back on wrist, cell phone turned back on, flat stone in one hand, heading back up the slope to the lane, I am diminished, specialized, narrowly focused, once again a member of the tribe, after a brief vacation of being released from it all.

Chapter 20

MIND AND MATTER

I was in my early thirties when RoseMary and I moved to Northern New Mexico. Before planting our first garden in the backyard of our rented adobe house in the summer of 1970, I had done very little physical labor. At age eight, I spent a painful afternoon hoeing weeds on the stony ground of a teacher friend of my parents in our hilly east San Diego suburb. In high school, I inherited from an older student a job as a gardener for an elderly doctor and his wife. Dr. Hurley wrote out afternoon instructions on a small spiral notepad, notable for such phrases as "Repair to the back yard and water the begonias." He was a slight man in his eighties, bony face often smeared with calamine lotion. Outdoors he wore a tan pith helmet, khaki shirt and pants, and walked with a shuffling gait, large hearing aid whistling with feedback from a chest pocket. When he didn't feel well enough to come outdoors in the afternoon, his severe, buxom wife handed me the instructions out the back door.

Besides watering the begonias, the front-yard bananas, and the backyard avocados, I was asked now and then to wash his maroon 1940 Chevy sedan, a car then old enough for me to dream of eventually coveting. I swept the walkways and the cement drive that sloped down to Lemon Avenue, mowed the lawn between the twin palms that presided over the front of the rose-colored stucco house and its red tile roof. At the end of my junior year, I announced that I would be attending the University of Chicago as an early entrant. The Hurleys were appalled. That communist university?

Aside from dragging garden hoses around, the work was hardly physical. I was skinny and tall, so I chose military science in high school instead of gym, becoming a member of the California Cadet Corps, the state version of ROTC. There were regular marching drills with heavy war-surplus bolt-action rifles. I don't recall any of these ever being fired. We were once taken to Camp Pendleton to watch Marine Corps demonstrations of howitzers and mortars. That was when I decided the military was far too noisy for my delicate ears. Classes consisted of overblown lectures on military discipline, the caring of uniforms and insignia, the correct way to fold hospital corners when making a bed. Under the tutelage of those seniors who had attained the ranks of officers, the real lesson was how to swear. Military science was guns, uniforms, hospital corners, and swearing.

A fond memory remains. One summer several hundred of us uniformed cadets were loaded up into an old troop train—the windows actually opened—at Union Station in San Diego and transported up the coast to the Signal Corps base above San Luis Obispo for a two-week camp. The base was located just below the famous Southern Pacific horseshoe curve on the

Los-Angeles–San Francisco line. During two a.m. guard duty, we were treated each night to the steam engine and passenger coaches of the Coast Starlight, brightly clad in silver and orange, rumbling and screeching down the spectacular curve on the way south to LA.

In my teenage years I played no sports. I had a weak sense of my own bodily presence. An awakening was to come on Lesbos when I became friends with a young French painter, François Fort, who wore his smooth body, clothed or unclothed, as an attractive item of his wardrobe. He was the first young man I knew who had let his hair grow down to his shoulders. He claimed he had once been kept by a gay French movie star, but he bedded down an impressive number of women who visited Molyvos in the summer of 1964. We met up at the shingle beach most afternoons, often had dinner together with the other expats, and were generally inseparable, except when he was courting the latest new female arrival.

I would never acquire his bodily pride, but years later, when RoseMary and I set about making the adobe bricks for the first two rooms of our house, I became addicted to moving rocks and lumber and beams. A fully laden truck gave me a sense of radiant physical accomplishment. Building the house along with gardening came to seem the natural first steps to taking up farming. And to the natural-seeming alternation between summers of the body and winters of the mind.

Yet I later saw this as a simplification. In the summer, out in the field, on the tractor, hitching up and unhitching implements, and in the work of repairing and adjusting equipment and tools, there are many details I take for granted, many minor techniques. Even though I now delegate much of the heavy lifting to young workers and interns, I am often struck by how the obvious to me—in hitching the rototiller to the tractor, say—remains still something of a puzzle to minds new to the work. A factor here is sheer repetition. After you have done something fifty or a hundred or more times, the sequences and patterns of gestures slip away into the back of the mind, becoming like the supposedly never-forgotten ability to ride a bicycle.

In physicality, mind is still there even if seeming to be absent. Just as, deep in some mental trance while daydreaming or writing or painting, the body still prompts us to eat, drink, stretch.

Chapter 21

THE HUMAN COMEDY

We sell most of our garlic and shallots at the Santa Fe Farmers' Market, Tuesdays and Saturdays. Like all public markets, it is good for people-watching. From behind the counters displaying our garlic and my books, with a folding screen to one side on which are hung decorative garlic bunches, I scan the crowds for the familiar faces of old friends and customers, who are likely to stop and say hello even though they may buy nothing on any particular morning. I'll even hail down those I want to talk to if it looks like they may stroll unseeing past the stand.

Older customers who know exactly what they want to buy and from whom are the first to arrive. From eight to nine the crowd thickens. Tourists and customers will add to the ten a.m. rush. The 10:30 Rail Runner commuter train from Albuquerque, whose tracks make up the western border of the market, will disgorge a new flood of people. Occasionally I am conscious of not having seen a regular customer who was showing signs of rapid aging or illness the season before. They now may be too ill to attend. Or may have even left this world. I know most regular customers by only their first names, though some names cycle in and out of memory. Is it "Michelle" or is it "Melissa"? Is "Jim" or "Joe"? I must have hundreds of names in my mental address book, including fellow farmers and spouses I have known for decades.

Customers stroll by, the thin, the plump, the fat, the obese, young and old, the rich, the poor, the homeless, Anglos, Chicanos, Asians, Blacks, Native Americans, old hippies, spangled and tattooed eccentrics of all ages. Sometimes I try to imagine how older faces looked when young, before the wear of age, before the sun blotched, before alcohol puffed, before years of anger or sadness twisted, before disease emaciated, before time roughened and hollowed and swelled the features into what often seem caricatures of the once svelte, the once smooth, the once open and radiant. The flowing crowd is like a river of time, though not in linear sequence. It is random, often surprising. Now and then I glimpse younger and older versions of long-absent friends. Many in the crowd are indifferent to what is going on around them as they chew on a sweet roll or burrito, sip from a bottle of water or a cup of coffee. Others are loaded down with bulging shopping bags. There are those who have just discovered the market, exude enthusiasm at everything they see. Many of my customers have read at least one of my books, inquire about what's coming next. Others express surprise that as a farmer I write books—as if I have crossed some vague class line.

At the end of the day, I judge the market as a good one when I have sold as much or more than I expected—and, more important, when I've had a good conversation with an old friend or with a new customer or with someone who has read one of my books or with someone who has passed on some interesting information about growing garlic or about a connection that might help with our ongoing battle with the Chinese garlic importer.

The Santa Fe Farmers' Market may not be of the philosophical caliber of Plato's Agora. But a marketplace is better when it's not just about business, and business is better when it is about more than just business, when it's also about making connections of a philosophical, social, or political nature.

"Only connect," advised E. M. Forster. Which, besides its locally grown fruits and vegetables, is what makes for the best farmers' markets: the degree and intensity of connections they inspire.

Chapter 22

THE PERFECT AND THE IMPERFECT

There is another consoling effect of farming. It has to do with notions of perfection and imperfection. In farming, in gardening, in working with the earth, in being immersed in a natural environment, you are in the presence of universal irregularity—of form, surface, texture, shape. You can call a garlic bulb "perfect." Or a tomato or an ear of corn or an autumn maple leaf or a sunset, cloud formation, a snow-covered landscape. But this is really to say that their irregularities are pleasingly arranged or displayed in reference to aesthetic ideals. I take pleasure in gravel surfaces underfoot, sections of irregular pavement, the jumble of river rock beneath the clear surface of the water, the rills and swirls and waves of the surface of a stream or river flow, the patterns of snow melting on pavement, the disintegrating mosaics of dry leaf litter along the edges of the lane. These are less appreciated, except by painters like Dubuffet, a favorite of mine.

Every garlic bulb and plant resembles its fellow cultivars, yet no two are identical: the leaves of one are nibbled, the bulb of another is asymmetrical, some bulbs are deeply striped in purple, others have thick long roots. In natural products we expect a range of variation, though commercial displays of fruits and vegetables seek to minimize this. The marked, the deformed, the blemished are culled out, even though many defects have no effect on taste. Russet apples with patches of rough brown skin are popular in France but not in the US.

Perhaps in these assessments we hold in mind an ideal bulb, a perfect bulb, a Platonic bulb. We have been conditioned to view people in the same way, in relation to movie stars and fashion models, though we know perfectly well that physical attractiveness or lack of it is often not connected to the qualities of character by which we judge others and choose our friends and mates. But these images of ideal female and male bodies infect how we view each other. Most of us are like russet apples, with less-than-ideal hair, blemished skin, oddly formed heads and limbs. That is, after we have left the blush of youth.

Yet paradoxically we celebrate irregularity in painting and sculpture and photography, to the point of grotesqueness and even beyond. Our values are divided, contradictory. This is exemplified in the case of industrial products, from the custom-made to the mass produced. We expect and indeed demand perfection in the panoply of tools and utensils and clothing and vehicles with which we manipulate and thread our way through our lives. The more we pay, the more perfection we expect, in smoothness and sheen of surface, uniformity of color and texture, amplitude of sound, sharpness of image, speed and smoothness of operation.

Many industrial products set impossible aesthetic standards for their all-too-human, all-too-natural users.

Industrial perfection surrounds us on all sides in the products we use throughout the day and night. The message they convey to our subconscious is that we, as humans, are flawed, imperfect, unreliable, dangerous, disorderly, and riven by irrational fears and desires. I sometimes think this manifests itself in my daily life when I become irritated at the recurring messiness of the kitchen or the area around the fireplace or how quickly the floors become dirty, and when I become exasperated at the very fact of dirt and dirtiness itself.

Yet in better moods I can fight back. My always imperfect garlic is produced with a minimum of environmental desecration. When I contemplate a perfectly formed wine bottle destined for the recycling bin or a toaster with a burned-out element or an LED bulb dead for some mysterious reason, I am struck by how the complex industrial processes that manufacture them have failed to deal with these almost-perfect objects in an environmentally responsible way. Given the fact that most products of our vast mining and manufacturing processes end up in landfills around the world, this form of perfection turns out to be a curse, whose effects will transform and possibly destroy the planet itself. When I ponder this, then I can celebrate who I am—as an exemplar of imperfection, along with my fields and their plants, and the very earth from which we all spring.

But perfection, fine. That is, if in the realm of most manufactured objects, single-use would be prohibited. It should be required of us to return the empty glass bottle before we are allowed by buy a new one. And the empty carton, the empty can, the empty plastic container, the empty pallet, worn-out jeans and jackets, broken toasters, light bulbs, electric and gas appliances and tools, and automobiles. As things now stand, manufacturers and retailers profit by passing on to individuals and municipalities the cost of recycling and disposing of consumer products—and by burdening the environment with the pollution generated in long-distance transportation and recycling processes and from landfills and the megatons of waste that end up in rivers and the oceans. One of the benefits of a "return and reuse" program would be an increase in local employment, a throwback to a time when, for example, every city of even modest size had its own 7UP and Coca-Cola bottling plants filling up refillable glass bottles.

Chapter 23

SHOT ACROSS THE BOW

On a Sunday afternoon in January 2016, an older red Honda Civic rattled up the drive. Alerted by the dogs, I stepped outside and was greeted by a young fellow who emerged from the car and handed me a thick manila envelope. He said good-bye and drove off. I had been served with a summons. This was nothing like the Hollywood versions of the event in which someone bumps into you in a crowded elevator and slaps an envelope to your chest or a cyclist crashes at your feet and as you help him up he thrusts a summons into your hand, that sort of thing. Two days later the same process server returned with a revised version of the summons. I asked him about the job. "The pay is great," he said. "They pay by the mile." He had driven up from Albuquerque, a two-hundred-mile round trip.

On the Sunday afternoon in question, I sat down in my armchair in the living room and opened the envelope and began reading through the seventy-four-page summons. It was filed in US District Court in Los Angeles by Winston & Strawn (offices in North America, Europe, and Asia) on behalf of Harmoni Spice International. I was one of twenty-one defendants. We were collectively accused of one thing after another, the crowning charges being extortion and racketeering. A fellow garlic grower I had recently invited into the Commerce review process (and who will remain nameless for reasons that will become clear later), Ted Hume, and a young attorney Ted had recently hired out of law school were among the defendants, along with Ruopeng Wang and Jack Bai. I was unfamiliar with the names of the remaining Chinese defendants, though it's possible I had met them on the previous summer's trip.

Halfway through reading the summons, I thought the whole thing was crazy: a shotgun blast of allegations in the hope that one or more might hit the target. The next morning, I chanced upon a friend who was the daughter of a lawyer. She guessed that the suit was a "strategic lawsuit against public participation," or a SLAPP suit, intended to intimidate and silence those who might dare to challenge the plaintiff by exercising their constitutional right to petition the government. Events soon confirmed this. I was reading an email from Ted, saying that he'd found a small law firm in Orange County, California, to represent us when the phone rang. An elderly male voice introduced himself as George Mastoris. I later found his name on the summons letterhead: he was one of the New York firm's partners. He asked whether I had yet obtained legal representation.

"Yes," I replied, "though I'm not certain of their names. But wait a minute," I added, realizing I had the information right in front of me. I passed it on to him. End of conversation.

On reflection I thought it odd that a senior partner in this large law firm would call me, not a secretary or a legal assistant or an office manager or any one of dozens of junior attorneys. The purpose was clear. Mr. Mastoris wanted to hear what I sounded like. He wanted to hear how articulate I was. He wanted to hear whether I might be an uneducated farmer of halting speech. He no doubt wanted to hear whether I was frightened or intimidated. I disappointed him on all accounts.

Our new lawyers, Tony Lanza and Brodie Smith of Lanza and Smith in Irvine, California, were surprised at how quickly Harmoni offered to settle, a mere four or five days. The offer was simple: they would drop the case if we withdrew our request for administrative review of Harmoni's garlic importing business with the Department of Commerce. They would also help us extricate ourselves from the RICO, or racketeering, charge. Gosh, thanks, but no thanks. . . .

A positive effect of the suit was that it seemed to fully engage my fellow farmer in the case. He was the owner of a farm fifteen miles away and at an altitude one thousand feet higher than my place. At my invitation, he'd joined the review request with Commerce back in October. I didn't know him well at the time. We had booths diagonally across from each other at the Santa Fe Farmers' Market. I had sold him a Dodge van a couple of years before. In recruiting him to the cause, I'd explained that the object of the review process was to level the playing field in such a way that the price of imported Chinese garlic would rise significantly. That was primary. Secondary was leveling the playing field among all importers by ending Harmoni's long-term advantage given by its rubber-stamped zero rate of duty, which had enabled it to obtain a monopoly on imported Chinese garlic. I suggested that if we were successful, we might well be rewarded financially, but this was not a promise. He joined, but I sensed some reluctance in his participation—until the lawsuit, whereupon he set about researching the whole garlic importing business. He soon knew more than I did, and both Ted and I learned much from his research. He found the lawsuit personally insulting. But it also distanced the possibility of any financial rewards.

For Ted, a racketeering suit over a dumping issue was "unprecedented." He had never seen this before. This was a civil suit; the RICO statutes were mainly intended to deal with criminal issues, not civil. Unusual also was the fact that Ted, as my attorney, was named as a defendant.

Harmoni hoped the suit would make us fold. On the contrary. In addition to this miscalculation, their fancy international law firm had made an interesting mistake. SLAPP suits are illegal in California, where the suit was filed. Most of the suit was eventually dismissed. Harmoni was required to pay our legal costs. Later the racketeering charges against US defendants were dismissed "with prejudice," meaning that the charges couldn't be warmed over and resubmitted.

A friend later described a firm that engages in these kinds of harassing practices as a "gutter law firm."

Not to be deterred, Harmoni appealed the racketeering decision.

Chapter 24

A LEGAL LIFE

Before the Harmoni lawsuit and its subsequent actions, my view of the legal system had been singularly positive. Back in the eighties, I was on the board of Northern New Mexico Legal Services (as it was then known), which served low-income people on issues of public benefits, domestic relations, and water law. My low income qualified me as a "client eligible" board member. The rest of the board members were Hispanic, Native American, and Anglo attorneys in private practice. During the Reagan years, I came under suspicion by the federal National Legal Services Corporation, funder of the state and local programs, for being a published author. As everyone knows, published authors are all well off. If only . . .

The legal services board was a congenial group. Eventually I became board chair. The work of the board was to set priorities and deal with staffing issues in the far-flung offices—Santa Fe, Taos, Las Vegas (NM), and Gallup. One of our annual meetings was held in Gallup, where the two attorneys stationed there reported how challenging and fascinating it was to establish the jurisdiction of potential clients. They often had to drive out into the country to determine whether clients lived on private land, New Mexico state land, or Navajo Nation land, in which case they should use Navajo Nation—Diné—Legal Services.

An urgent issue when I first joined the board was whether we could represent Hispanic and Anglo defendants in the very long-running Aamodt water adjudication suit, the purpose of which was to formally establish individual water rights for three Native American Pueblos (the plaintiffs) and the non-Indian defendants. The US Forest Service and the Bureau of Land Management were also involved. The Pueblos and Federal agencies were represented by the US Attorney's Office. But Legal Services policy dictated against entering into suits with low-income people on both sides. The dilemma was eventually resolved when Congress appropriated funds to assist Hispanic and Anglo defendants.

My good impression of the legal system was enhanced by serving on a jury for an assault-and-battery case at the Rio Arriba County Courthouse in Tierra Amarilla, site of the famous courthouse raid by Chicano land-rights activists back in 1968. I had read about the raid in the newspapers while traveling across the country on our way from Dublin and New York to San Francisco, never dreaming we would end up a year later in the very county of the famous, or infamous, standoff over Spanish land-grant rights. In any case, during the jury-selection process, I complained to Judge Timothy Garcia that I had a two-hundred-mile round trip between my house and Tierra Amarilla. With good humor, he sympathized but pointed out that he had a similarly long commute from Albuquerque.

The trial was complex, fascinating. There had been a party in Española, drinking. A drunken guest had crashed onto a bed that happened to be occupied by the host's passed-out wife. A fight broke out. A pickup was trashed. Guilt seemed widely distributed, if a little unevenly. The prosecuting attorneys were good, but the public defender was of star quality. Trial over, the jury adjourned to the jury room. I was the only Anglo. I knew what would happen next: I would be elected foreman. But then, after a knock on the door, the bailiff entered and announced that the judge had selected me as an alternate juror, excused unless and until called back. As I was leaving, Judge Garcia called me over and suggested I might want to comment to the attorneys about their performance. I congratulated them all. It was a complex case. I hoped a decision would be reached that tempered justice with mercy in a way that didn't ruin the defendant's position as the family breadwinner. A neighbor acquaintance on the jury later reported that indeed such was the outcome.

The legal system I participated in back in the eighties was not the one of huge international law firms with deep-pocket clients that I was slowly dragged into starting in 2014.

Friends occasionally comment that I don't seem particularly agitated over being at the center of all these legal battles. Other than regularly solving the problems of the world during brief strategy sessions at three a.m., I rarely lose sleep. And I suppose I find interesting what many people would agonize over. Each new turn of our battles with Commerce and Harmoni reveals a new aspect of the world I would otherwise never have learned about first hand, particularly the ways that rich and powerful corporations and their enabling law firms and government agencies can distort the laws to better serve their corporate interests.

There is, of course, much to be angry about—the vast amount of time and taxpayer money and legal resources being wasted over an administrative process that should have been quick and direct, the personal attacks that Ted and I have been subject to. But if I have learned anything in this long life, it is that anger can lure you into a cave of blindness and ignorance from which it is often difficult to find your way out.

I haven't been able to turn the other cheek entirely, but I have tried to remain calm and thoughtful and find meaning wherever I can, even in the murky legal depths.

Chapter 25

MONEY

Ted's legal experience was, predictably, of a different order than my own. After graduating from Yale, where he was a fraternity brother of the younger George Bush, he obtained an MA from Georgetown and a law degree from American University. He served in several high-level positions with the US Treasury, the Customs Service, and the Office of US Trade, then worked for Texas Instruments and a Hawaii law firm before forming his own firm specializing in international trade law. For the past several decades he has represented Chinese exporters of hand tools, lug nuts, wooden bedroom furniture—and garlic. Much of this work consists of conducting "verifications" of Chinese producers in collaboration with Department of Commerce officials in order to establish production costs, which are used to set rates of anti-dumping duties. This is relatively dry work devoid of the human messiness that other areas of the law grapple with. Garlic, after all, is a boring commodity. How could it possibly set off an unprecedented legal war?

Easy: money. With each costly legal move against us, it became clearer that something very large was at stake for Harmoni and its allies, though exactly what was to remain unclear for a long time.

And, as Ted observed to me, the whole system had become unrecognizable. The anti-dumping process was now dominated by large corporations served by large corporate law firms. Ted, as a sole practitioner representing small farmers, had become an anachronism—or worse, an irritant with the potential to upset the profitable corporate apple cart involving other industries that had learned how to collaborate with Chinese importers to game the anti-dumping system to their advantage, work that employed countless well-paid trade lawyers.

What had started out as a simple administrative request to the Department of Commerce had turned into the much larger issue of how to reform a government system that had been essentially corrupted by the importers it was supposed to regulate. From the beginning, I think Ted sensed that it was probably a David-and-Goliath situation, but it had become clear that he had misjudged the size and determination of the Goliath.

Chapter 26

A PALL

From the beginning of 2016, when the lawsuit against us was filed, an air of unreality hung over the farm, not so much a cloud as a kind of haze like that generated by a distant forest fire. As the year advanced, I continued to go through the motions of irrigating and weeding and harvesting and selling garlic at the farmers' market, paying bills and updating accounts, but with a sense that in some oblique way my actions were being observed from afar—that if I was not actually being spied on, everything I did or said would be the subject of hostile scrutiny sooner or later.

It was an odd feeling to know that a small army of attorneys, including Department of Commerce staff members, would go over every word of my public filings. There were bound to be discrepancies and inconsistencies in my statements. Would these be blown up into allegations of deception and even fraud? Indeed, over time, scrutiny was to be applied far more exactingly to my small garlic operation than to the large corporate network of garlic suppliers to Zhengzhou Harmoni Spice (China) or to the relations between Harmoni International Spice (US) and the four members of the California-based Fresh Garlic Producers Association (Christopher Ranch, Vesey and Company, the Garlic Company, and Valley Garlic), which we believed were the major conduits for the distribution of Harmoni-imported garlic in the US.

Sometimes I wondered what the legal interns and attorneys of Harmoni's three international law firms thought about when beating up on a small New Mexico garlic producer. Did they feel any remorse, any pangs of conscience, any doubts about their cause? They must be bright and intelligent attorneys, products of the best law schools in the country. But of course they knew perfectly well what they were about. They might even be happy to rack up expensive billable hours studying Ted's and my declarations with Commerce and researching the minutiae of administrative and court decisions concerning anti-dumping statutes.

Should I take seriously concerns by a few friends that I might be in physical danger? Was I doing anything wrong in the way I was running the farm? How could there be any question that I grew garlic for a living? I harvested garlic, I sold garlic publicly, I planted garlic, processes I had already described in a book published by the University of New Mexico Press and in countless columns for periodicals in Santa Fe and Albuquerque.

I was in business. The farm was a commercial operation. It was small, but no "hobby" farm: I filed IRS Schedule Fs, employee records, depreciation and insurance records, and I obtained business licenses. I have always kept good financial records, which among other things are needed for loan applications. I also believe that when self-employed people fail to report cash income, they are in effect thrusting more of a tax burden on wage earners who have no way to conceal income.

A sense of impending scrutiny led me to begin recording each individual garlic sale, no matter how small, in order to establish the weight of the entire crop at the end of the season. Since we harvest the whole garlic plant, we can only estimate the total weight at the beginning of the season. Until recently I haven't bothered to record individual sales in order to calculate a more exact weight. True, during busy moments at the farmers' market, we occasionally fail to record a small sale. And now and then someone will drop by the farm and buy a pound of garlic, which I'll neglect to log. In all, perhaps this is a matter of a hundred dollars over the course of the summer, or some eight to ten pounds of garlic.

In 2017, we sold about 1,600 pounds of garlic and planted another 700 pounds, for a total of 2300 pounds grown. This is not a good ratio of planting weight to harvest weight. It would be much better in a warmer climate, such as the Central Valley of California. But more important than quantity is quality. In terms of flavor, we can claim excellence.

Through the farmers' market, I consider face-to-face encounters to be the norm. By contrast, it's disconcerting to read pages of allegations against me and my farm without being able to confront my accusers face-to-face.

There is also a side effect: when words are committed to print, in physical or digital form, they become set, seemingly immutable. Public records last forever. Or for enough time to seem forever.

But such is also the case with essays and books. In the past my writing has served as a means for me to comprehend my world, not to justify, not to set the record straight.

But guess what: here I am. At least in part. Though also I hope to discover some new meanings in the four-year legal and administrative saga over Chinese garlic through the writing of this account.

Chapter 27

EL BOSQUE GARLIC FARM GOES TO WASHINGTON

In late May of 2016, Ted arranged for an ex parte meeting with the Department of Commerce staff in the Herbert C. Hoover Federal Building on Constitution Avenue. The object of our meeting would be to give dimension to our legal filings, to convince Commerce officials that we were real people with a real issue. "Ex parte" means that our meeting would be without representatives from Harmoni or anyone else opposing our request for review. Ted and I, along with my fellow garlic grower, would fly to DC late Wednesday, spend the night, meet with Commerce late Thursday morning, and fly home in the afternoon just before the Memorial Day weekend.

Wednesday evening after dinner at our downtown hotel, Ted retired to prepare remarks for the morrow while my neighbor and I walked over to the White House and back, then agreed to hike around town the next morning before the 11:00 a.m. meeting. Our morning destination became the Vietnam Memorial. In the course of our long walk, we did what new acquaintances often do: we summarized our lives and compared notes. Though we had known each other for years as competing garlic growers, we had never had this kind of exchange. It turned out that he had also spent time in Molyvos some years after my summer in the village. There were other points of similarity despite our twenty-year age difference. By the end of the walk, I felt I had acquired a reliable new friend. Not someone I was likely to become much closer to, but who I could count on in the daily business of farming. I felt he was a solid ally in the complex battle with Harmoni.

The Vietnam Memorial was marred by the presence of herds of chattering high schoolers who no doubt only had each other in mind, not the war that traumatized their parents' and grandparents' generations. Not then understanding the order of the names inscribed on the wall, I was unable to find the name of an army medic, Michael Valdez, who was posthumously awarded the Medal of Honor. He grew up in my New Mexico village, and I had known his late father well.

The Hoover building was close enough to the hotel that we were able to wheel our luggage down the gentle slope in the direction of the Mall. We were permitted to keep our bags with us after going through security. The Commerce building is a gray neoclassical stone pile with interminable hallways lined with wooden doors opening into offices; it dates from 1932. A staffer guided us to a small, windowless conference room with a long oval table, where we were joined by seven Commerce attorneys and policy staff members, presided over by the Deputy Assistant Secretary of Commerce, Christian Marsh. Of the group, Marsh was the only political appointee. Ted knew many of the group, having met with them multiple times before over various anti-dumping matters.

Marsh was very welcoming. He was an attractive, sandy-haired fellow in his fifties. Ted started out by saying why we were here and what our issues were. Our "standing" as commercial garlic producers, in the case of my neighbor and myself, was the most important: it was key to Commerce eventually agreeing to conduct a review of Harmoni. This is why Harmoni had spent so much time and effort questioning our standing, in the hope that Commerce would eventually accept Harmoni's contention that we were mere hobby farmers with no skin in the game.

My fellow grower went next, explaining eloquently that as an American citizen he felt his constitutional right to petition the government was being questioned. He talked about his farming practices. Indeed, in weathered jeans and a plaid shirt and no tie, and with calloused hands, he was clearly someone who worked with the soil. When my turn came, I slid a copy of my garlic book and the plastic sack of freshly pulled Turban Chengdu bulbs across the table to Undersecretary Marsh.

"My wife will be delighted!" he exclaimed, holding up the garlic.

There followed a series of questions and comments, the sort of exchange Ted later said was unprecedented in his experience with these hearings. Normally ex parte hearings are attended only by attorneys who quibble over the fine points of anti-dumping statutes and legal precedents. But my neighbor and I were "real people" who worked with our hands for a living. It was our businesses that the anti-dumping laws were presumably written to protect.

Throughout the meeting, which lasted much longer than the scheduled half-hour, a young attorney at the far end of the table maintained a warm smile throughout our presentations. We later learned that she had grown up on an Idaho farm.

At the end of the meeting, Marsh escorted us out and showed us a shortcut back to the street. Once outside, Ted said, "I've met with Commerce dozens of times, and that was the best meeting ever."

Very satisfied with our work, we flew home—mesmerized by an Ethiopian-born flight attendant who was a President Obama look-alike.

Chapter 28

UNPRECEDENTED

In June of 2016, the Department of Commerce accepted our standing as a producer of garlic, a "like product," which meant that Commerce's administrative review of Harmoni had the green light to proceed. During the prior six months, Harmoni had fought us tooth and nail, down to the definitions of individual terms, such as "like product" and "wholesale." In response to the Commerce decision, Harmoni was required to answer a detailed questionnaire about its complex Chinese operation, which Ted believed involved some thirty suppliers of garlic for its export business. The questionnaire, in effect an audit, covered all aspects of their business in China and that of its suppliers: production costs, including labor, utilities, shipping, et al. For as large an operation as Harmoni had become, this was burdensome but certainly something they should have been prepared for, especially given the fact that for the past dozen years Harmoni itself had actually requested that it be reviewed, a request echoed by the Fresh Garlic Producers Association of California. But, like clockwork, the requests had been withdrawn at the same time by both entities by the annual deadline. In a feeble attempt to mask their ploy, they also "withdrew" a number of importers who were listed but in fact were not importing for the years in question.

It took me a long time to understand quite what was going on here. The reason that Harmoni (US) and the Fresh Garlic Producers Association requested a review of Harmoni each year was in fact quite simple: it inserted them into the review process, giving them standing to oppose anyone else, such as me, who actually wanted Harmoni to be reviewed. It also gave them the power to withdraw Harmoni from the review process, which they couldn't have done had they not filed review requests with Commerce in the first place. In other words, in order to be able to game the system, you have to enter the game, which they had now played for twelve years running, as of 2016.

The whole exercise was clearly a sham intended to protect the relationship between Harmoni and its California collaborators, who profited by distributing duty-free Harmoni garlic. The anti-dumping duty would serve to raise the price of Harmoni's imported garlic to a level closer to the price charged by US producers, ending the unfair advantage posed by low-priced "dumped" garlic. By all rights, of course, the Fresh Garlic Producers Association of California should have been on our side, seeking to end Harmoni's zero duty rate—but that would have ended a profitable relationship of its four members with Harmoni.

According to Ted, Harmoni's response was—you guessed it: unprecedented. It is normal for an importer to be granted one or two extensions of a week or two before responding to the Commerce questionnaire. Harmoni asked for and got eight extensions totaling nearly three months. Commerce finally denied a ninth extension request. Whereupon Harmoni walked away, refusing to answer the mandatory questionnaire.

Harmoni's attorney later admitted that this strange decision had been a mistake because it immediately subjected Harmoni to what is called the People's-Republic-of-China-wide rate, which is $4.71 per kilo. As a result, Harmoni owed the U. S. Treasury over $200 million in anti-dumping duty, a figure based on its imports for 2015 alone. Ironically, it was Harmoni that had helped establish the high China-wide rate some years before as a way to discourage competition from other Chinese exporters.

The Department of Commerce annual review process has two phases in the decision process. At the end of phase one, in early December, Commerce issues its preliminary finding. In our case, Commerce reaffirmed our standing as garlic producers in its December 2016 preliminary finding. It reaffirmed its decision that Harmoni was subject to $200 million-plus in anti-dumping duties, for failing to respond to the mandatory questionnaire. Commerce also pointed out that statutes did not cover our relations with other importers, which is to say that that our relations with other importers, in the form of discussions or financial transactions, were not relevant to the review process. That apparently applied both to us and to the tight financial relationship that we believed existed between Harmoni and its California garlic producer allies.

Not long after that, US Customs and Border Protection required Harmoni to post an enhanced bond to cover the duty owed. Harmoni contested that decision with the federal Court of International Trade in New York, which ruled against Harmoni, citing the fact that it had made a series of poor decisions in attempting to avoid the Commerce audit.

In May 2017, Commerce held a public hearing on its preliminary finding. According to Ted, it was a raucous shouting match among a room full of attorneys. Again, unprecedented. Unfortunately, a previous commitment prevented me from attending. The last phase of the administrative review process takes place annually in June, when Commerce makes its final determination.

Through the first half of 2017, it appeared very clear that we were winning on all fronts. Or rather, on all fronts but one . . .

Chapter 29

AN ASSORTMENT OF DEADLY SINS

Not long after our DC trip, I was somewhat troubled by a discovery about my co-filing garlic farmer neighbor and his wife. They had met with Ted and Renate to complain about their lack of money and about raising their two kids in a house with leaky roofs and no hot water and no indoor bathroom. Ted and Renate agreed to give them $5,000. Within a day or two of this, I had lunch with the couple in the local café. No mention was made of the gift. Several weeks elapsed before Ted and Renate told me about the request, which they had felt uneasy about.

Around the same time as the gift to the farmer couple, Ted's office manager borrowed $3,500 from Ted and Renate. She was deeply in debt. That both these requests for money were made within days of each other later made the Humes wonder whether the two parties had colluded behind their backs. A couple of months later, when the office manager's husband kicked her out of the house, Ted and Renate helped her out by putting her up in their guest house for ten weeks while she reorganized her life.

For reasons I have never understood, the garlic farmer couple were suspicious of Ted and Renate, despite my assurances that I trusted them and was very fond of them. One issue was that Ted represented other Chinese importers, including Jack Bai. My neighbor's contention was that we were only helping other large Chinese importers enter the market. Eventually, in the filings he would submit to Commerce against me and Ted, he would refer to us as "pawns." But over lunch and in later conversations, neither my garlic farmer neighbor nor his wife insisted on that argument. My response was, yes, if successful, our actions would benefit other Chinese garlic importers. But if Harmoni is finally obliged to pay duty, then the price of garlic in the US will rise, perhaps significantly, thereby benefiting American growers, including ourselves. Also, the extreme efforts Harmoni had gone to in fighting us made me wonder whether they might be hiding something truly serious.

As he had the summer before, that summer my fellow garlic grower borrowed my garlic undercutter, saving him huge amounts of hand labor in his very weedy garlic field.

In the fall, he and I met with Ted in Taos. Ted was heading off to China to meet with our allies there, his old friend Ruopeng Wang and exporter Jack Bai. Ted wanted to take a wish list of things that would help New Mexico growers increase production, with the thought that the stronger we were the more likely we would be to continue to file administrative review requests of Chinese garlic importers with Commerce. Ted and I turned the task over to my neighbor, who seemed anxious to undertake the assessment. A week later he came back with a $2.7 million budget that covered land and equipment and salaries for several years. The funds would be managed by an LLC, which we formed, with my neighbor as prospective manager. I was astounded at the boldness of the request but unfortunately did not object. Nor did Ted.

In November, we all met for dinner at Five Star Burger in Taos: Ted and Renate, the garlic grower and his wife, RoseMary and myself, and RoseMary's sister Zita. Ted and my neighbor and I sat at one end of the table, the women at the other. Ted reported that he had come back from China empty-handed. His Chinese friends felt we were not "real businessmen," and by implication we weren't worth funding for $2.7 million. My neighbor took umbrage at the description, but I couldn't help feeling that they were right. We were very small players, however pivotal we had become in the battle with Harmoni.

At the other end of the table Zita said something to the effect that it was nice that the payment I had received a year and a half before had enabled me to buy a new tractor and expand our photovoltaic array and buy some other equipment. That payment, the $50,000 gift from Ted, had come well before I had invited the neighbor into the review process with Commerce. My neighbor's wife misheard these remarks, imagining that I had been given this equipment over the past several months behind their backs without telling them.

I left dinner without realizing how seriously wrong things were going.

Chapter 30

EXTORTION? BLACKMAIL?

The very next day, my neighbor and his wife went to Ted and Renate's house and demanded that they pay them $25,000, failing which they would withdraw from the group and from the Commerce review process. The Humes gave them $10,000 as a loan, said the balance would be forthcoming by the fifteenth. Renate said it was a very unpleasant meeting. They later regretted paying them anything at all.

Learning of this later in the day, I emailed my neighbor and asked if we could meet immediately. I suggested that covering emergencies was what families were for. I believed his father was well-heeled, if not wealthy—a couple of years before, he had stayed for three weeks in our guest house, his large Lexus parked outside the Tower. I also thought there was money in his wife's family. He claimed he was too busy that day, but came down to the farm one afternoon a few days later. We spent an hour going over the situation. I explained that I had borrowed money to buy the tractor and the Polish garlic planter and had paid for the solar expansion outright. These were not gifts from our Chinese allies. I had the paperwork to prove it. It would take me awhile to dig it out of my files in my studio, but it was there. And all this was months before I had invited him to join the effort. He said he "sort of" believed me, a curious way to address my assertions. Well after he defected, I did receive other equipment from our Chinese allies.

In one of his emails he had repeated that if they didn't get the full amount, a "loan," from Ted and Renate, they would leave the group and retract their participation in the review process. And they wouldn't go "gently or quietly." I pointed out that most people would view this as extortion or blackmail. It's one thing to ask for money. It's quite another if the request is accompanied by threats of any kind.

We discussed these questions back and forth, though not heatedly. By the end of the hour, as I escorted him back to his pickup, he apologized again and again for what had happened.

We have never spoken since.

Chapter 31

THE BETRAYAL

A few weeks later, in mid-December 2016, my neighbor formally withdrew from the group and from the Commerce review process when Ted and Renate decided they shouldn't lend them any additional money, which they felt they couldn't afford. While I was teaching at Colorado College, my neighbor and his wife contacted a reporter for the daily New Mexican in Santa Fe. In their interview they explained how they had been "duped" by Ted and me and how the whole process was a "sham" to benefit Ted's Chinese clients. Ted was also interviewed and defended our position. When the reporter called me in Colorado Springs, I was suffering from the flu and could barely talk for coughing, though a positive line of mine was the last quote of the piece.

The article attracted the attention of the daughter of one of the Harmoni owners, who telephoned my neighbor's wife and offered to put them in contact with Harmoni's New York law firm. Legally the firm could not contact them directly, as the

Los Angeles litigation was still active. GDLSK flew them to New York and agreed to release them from the racketeering suit and to compensate them for "research" for an undisclosed rate, a fact which was stated in a public filing with the Department of Commerce. Shortly thereafter, my neighbor filed a twenty-one-page declaration with Commerce detailing how he believed he had been misled by Ted and me into participating in a "fraudulent scheme."

Some time before, the Humes had discovered that their former office manager had embezzled some $9,000 in credit-card charges and forged checks, for which they had fired her. They declined to press charges, at least until they discovered that Harmoni's lawyers had obtained confidential files, which could only have come from the fired office manager. Upon Ted's filing of charges in the Taos District Court, the former office manager counter-sued for malicious prosecution, using a high-profile Albuquerque firm. By her own admission in a formal legal declaration, Harmoni was paying her legal bills.

This was the second attack by Harmoni against Ted. GDLSK had also made a claim to the New Mexico Supreme Court that Ted was practicing law in New Mexico without a license. Ted is licensed in the District of Columbia, New York, Texas, and Hawaii, but not New Mexico. That case was dismissed on the grounds that no state license is needed to practice federal law. Once again, GDLSK, the international law firm, got its very basic facts wrong, as they had with the SLAPP suit with the US District Court in Los Angeles.

Ted was finally starting to fully comprehend the warnings he'd received a few years before, when he'd first moved to take on Harmoni. It was entirely clear now the lengths to which they would go.

Chapter 32

INTERNS

At the start of the new growing season over last couple of years, it's been a relief to put aside thoughts of legal machinations over Chinese garlic imports and begin thinking about the upcoming schedule of farm tasks and needs. For the 2017 season, I had already lined up a student from my December session at Colorado College. After a class in which I had alluded to the farm, student Ben Garinther approached me, asking whether I used interns. Ben was one of the better students in the group of twelve, a good writer and active participant in discussions, personable, tall, with a bushy black beard and curly black hair. We sealed the deal over a cup of coffee at Wooglin's Deli next to the campus in Colorado Springs. I decided he would fit in nicely with the farm scene, which turned out to be the case. He always answered with good humor RoseMary's daily question, "Who did you get your hair from?"

Ben was my third intern. The first was Jono Tosch, a student in the second spring fiction workshop I conducted at UMass/

Amherst for the master's program in 2011. He was a poet and an avid gardener. I didn't know how much I could teach him, but the modestly paid internship offered a pleasant summer in the Southwest. He was with us for the better part of two summers. He has gone on to become a professional house painter and general handyman, but still has time to write poetry and to assemble elaborate collages, which he displays in a couple of galleries. He regularly reports on the state of his productive front-yard garden in Northampton, Massachusetts, in which descendants of some of my garlic grow.

My second intern was Sam Tezak from Denver, a Colorado College student. During off-duty hours at the farm, he wrote two short novels before eventually turning to poetry, his major upon graduating. He went on to work for NPR in Washington as an intern fund-raiser and developer of stories for the news program before moving back to Denver to work for Outside Magazine. Since his summer with us, he has frequently driven down from Denver during visits home. RoseMary took to calling him "Denver Boy."

All three interns, plus two high school classmates of Ben's—Campbell Reid, who went on to work for a large farm in northern Mississippi, and Phil Babbitt, now in marketing in Austin—were with us for shorter periods during the summer of 2017. They worked on the farm during seasons without extended droughts or ditch-destroying flash floods or crippling infestations of insects. As workers, not farm owners, their labors usually ended at lunchtime or early afternoon, except on market Tuesdays and Saturdays with their four-a.m. wake up calls and twelve-hour days. The interns were not kept awake at night over worries about money, how to pay for manure and mulch, how to fit in maintenance work on the vehicles and machines, how to find the time to repair or replace some broken tool or part, or any of the other dozen small and large challenges facing a small farm.

What they learned, perhaps, was a love for working outdoors, weeding, irrigating, planting, harvesting, how to drive a tractor, and how to work some of the smaller machines. They learned how to pack the pickup for market and how to unpack it on returning home. They learned how to interact with customers at the market, how to sell produce. And I hope they acquired a sense of the place of a small market farm in the larger world of agriculture and trade—including, in the case of Ben and Campbell and Phil, the pivotal position of El Bosque Garlic Farm in relation to cheap Chinese garlic.

Chapter 33

TOOLS, TOOLS, AND MORE TOOLS

Late winter and early spring is also when I go through a mental inventory of tools to determine which need to be repaired or replaced, which need to be cleaned, oiled, sharpened, and what new tools I feel I need to acquire. Often I make lists of such things on a clipboard I keep next to my armchair. Of course, you never have all the right tools, particularly in an era when clever new devices appear on the market almost daily.

Minus those pruning shears and nut drivers and clippers that have disappeared into the bushes somewhere, my inventory of hand tools consists of the following: shovels, including standard long-handled shovels, a narrow planting shovel, broad-blade scooping shovels, two snow shovels, an ash shovel; four- and eight-tined pitchforks, garden forks; garden hoes, including scuffle hoes, narrow "collinear" hoes, narrow and broad ditching hoes, hand hoes or weeders; a grass rake and a garden rake; a dozen or more scissors and hand clippers; pruning shears; brooms, whisk brooms, dust pans, scrub brushes, wire brushes; gas and electric chain saws, a JawSaw for pruning, hand pruning saws, pole saws, a Sawzall, a Skilsaw, a jigsaw, four types of hand saws; claw hammers, ball peen hammers, sledge hammers, hide and rubber mallets, a tack hammer, a maul; wood and metal punches, awls, chisels, wedges; regular pliers, needle-nose pliers, diagonal pliers, locking pliers, Vise-Grips; open-end wrenches, ratcheting wrenches, pipe wrenches, filter wrenches, socket wrenches, nut drivers, adjustable wrenches; screwdrivers with flat, Phillips, and Torx blades; telescoping grippers; staplers; paintbrushes and scrapers; hydraulic, scissor, and floor jacks, jack stands; axes, hatchets; aircraft cutters, tin snips, utility knives, pocket knives, razor-blade cutters; step ladder, 6-foot ladder, 16-foot extendable ladder, scaffolding; rulers, tape measures, yard sticks, T-squares, levels; sprinklers, garden hoses, spray nozzles; a fence-post driver; a grinder; a drill press and hand power drills; shop lights and flashlights, extension cords; an air compressor; pry bars, a crowbar, small bars for extracting nails; a laptop, a tablet, a smartphone.

And a pair of five-fingered hands with opposable thumbs, the tool of tools, though sometimes toes and feet and even tongue and lips come in handy. All, of course, under the guidance of mind and imagination, which tell you which tool to use (or indeed invent) and how and when to use it.

Chapter 34

THE SUN ALWAYS RISES

Mind and hand create and guide tools, but it is easy to forget that the source of all energy, mental, physical, mechanical, electrical, is the sun.

My farm runs on photosynthesis, the process by which sunlight converts carbon and water into carbohydrates, into green plants, trees, seaweed. Petroleum and coal are the antique residues of the direct and indirect products of photosynthesis. In short, everything we do is ultimately solar powered.

I have been obsessed with what you might call direct solar power since the 1970s, when I was introduced to contemporary solar systems for generating heat, in particular, by the late Los Alamos engineer, Benjamin ("Buck") T. Rogers, who lived in the neighboring community of Rinconada. He helped me design my first Trombe wall in the late 1970s, consisting of glazing placed vertically a few inches out from a 12-inch-thick adobe wall painted black: in effect, a very thin greenhouse. The wall collects heat, which gradually moves through the mass and radiates out into the interior space over the course of the night. We now have two Trombe walls on our south-facing living room adobe wall, and a third on the south-facing adobe wall of the guest-house bathroom.

In the 1970s we also built our first "attached solar greenhouse" along the south-facing 18-inch-thick adobe wall of our new bedrooms. Two sets of French doors opened into the space. Following the fashion of the time, developed by the late Bill Yanda and his wife, Susan, the translucent fiberglass glazing was angled at 45 degrees. Unfortunately, this reduced headspace in the 8-by-30-foot space, and the greenhouse overheated from late spring until early fall.

Within a few years I corrected these problems in a second iteration. I made the south-facing glazed walls vertical and covered 4 feet of the 8-foot wide roof with a solid structure, which provided shade as the sun moved north in the summer. We also replaced the fiberglass with double-pane glass, using double-wall polycarbonate panels for the glazed portion of the ceiling. During our years of peak farm production, we started some ten thousand plants in the greenhouse, which also provided heat (and a bit too much humidity) for the house during the winter months.

Around 2013, the limitations of the 8-foot wide space—which we also used for curing up to a thousand pounds of winter squash in October—were becoming evident. We tore down the structure and expanded it to 14 by 36 feet, poured a concrete floor over insulation, and installed doors at both ends to facilitate ventilation in the summer. It turned out that the space was large enough to house a heat-pump water heater, which works the opposite of a refrigerator: it pulls heat out of the air to heat water. (A 10-by-10-space is nominally required). A heat-pump water heater in a solar-heated greenhouse is in effect a solar water heater, but minus the pumps and antifreeze, and attendant complexity and expense, needed for a rooftop solar heater in cold climates. Yet the system has a direct solar component in the form of two solar batch heaters mounted on the greenhouse roof. These are 30-gallon tanks enclosed in double-glazed boxes, a gift from a friend who removed them from his Santa Fe house. We are too far north for them to generate more than warm water in the winter, but they serve well as preheaters for the heat-pump heater, which paid for itself in propane savings in about a year.

The new greenhouse roof, the expanded solid portion, enabled us to add to the photovoltaic system a local company had installed on the Tower roof a couple of years before, giving us enough photovoltaic power to take care of most of our electrical needs, including the daily charging of first a 2012 Chevy Volt and then a 2017 model. In effect between a third and a half of my driving of the Volt is powered by the sun.

My hope is that these efforts will reduce my carbon footprint to minimal levels. It has been argued that the only real solution to the problem of global warming is through massive public policy changes, including a significant carbon tax, the installation of large-scale solar arrays, programs to encourage the conversion of all vehicles to electricity, and so on. But the vast infrastructure of fossil fuels, including still ongoing investment in it, will be a slow one to convert. What many individuals can do immediately is to install rooftop solar and switch to electric vehicles. Financially this will seem unaffordable to many—as in some ways it still is to myself. It has been argued also that the problem with rooftop solar is that in effect you are paying for your electricity twenty years in advance. Yet, by the same token, the environment can no longer afford for anyone to keep spewing carbon into the atmosphere at current levels, particularly when the means exist to minimize such emissions at the household level right now.

But best of all, once the one-time capital investment is made, solar energy is free for the asking.

Chapter 35

EL BOSQUE GARLIC FARM BECOMES A MOVIE STAR

In the summer of 2016, we were contacted by a New York production company, Zero Point Zero, best known for producing the late Anthony Bourdain's travel and food series. They expressed interest in doing an episode about garlic for a documentary series tentatively entitled *Food Crimes*. During the June garlic harvest, they sent out a videographer, who spent most of the day filming the various aspects of the harvest. The purpose of the shoot was to give Zero Point Zero ideas for future visuals. During the months following, the project waxed hot and cold. Finally they announced their plan to come out and film in early April 2017.

Following a writing conference in DC, I took the train up to New York, where I stayed with old friends on West Twenty-Third Street, not far from the recent trash-can bombing. Zero Point Zero's offices were a couple of blocks uptown, on Sixth Avenue.

I spent an hour with four staff members going over the Harmoni situation. I was impressed at how well they had mastered the intricacies of anti-dumping law. They were particularly interested in the history of my relationship with the other garlic grower and his wife and the steps by which it had become toxic. I was not to make this publicly known, but the six-part series was to be released by Netflix in early January 2018. Zero Point Zero had already completed an episode on Atlantic fishing quotas. The other episodes would cover honey, chicken, and nut and dairy allergies.

As arranged by phone, the five-member crew arrived at the farm at 5:30 a.m. on an early-April Saturday, took shots of me drinking coffee in the kitchen and driving away in the dark in my Ford pickup. Two of the crew accompanied me on the drive, shooting as we went. They continued filming as I arrived at the Santa Fe Farmers' Market, set up the booth, sold to customers, and packed up at the end. On returning to the farm in the afternoon, they took shots of the foot-tall garlic in the field before heading up to their Taos motel.

The crew returned at eight Sunday morning and spent several hours setting up in the living room, adjusting lights and sound-recording equipment, moving objects they felt wouldn't go well in the picture. They had me sit on the sofa next to the fireplace, where they interviewed me for two hours. I also read aloud passages they had selected from *A Garlic Testament*.

Earlier I had said I would be uncomfortable talking about my fellow garlic grower, to which the director responded that if I didn't say anything, someone else would.

Throughout, Ted sat waiting for his turn in the kitchen, while Renate took care of RoseMary in the bedroom. Her Alzheimer's was sufficiently advanced that I felt it important not to film her, at least at any length. After lunch, Ted was interviewed sitting at our kitchen table—another two-hour session, plus set-up time.

Later we had dinner with the crew at the Ranchos Trading Post in Taos. They talked briefly about interviewing a Chinese-American importer back east who had footage of Chinese prisoners peeling garlic for Harmoni. We were aware of the allegation and had seen grainy black-and-white photos, but these did not seem to constitute surefire proof. The crew's next stop was China, where they would interview Ruopeng Wang and Jack Bai, and where they hoped to look into the prison-labor situation.

The crew expected to have the project wrapped up by October, whereupon Netflix would go over the version and oversee translation into other languages, for a potential global audience of a hundred and ten million.

Chapter 36

BAD BEHAVIOR

I have been fortunate to have suffered betrayals at the hands of presumed friends only one other time. Or something close to betrayal—whatever you want to call finding yourself turned against in a vicious manner.

I became acquainted with an English novelist in the winter of 1988 at the MacDowell Colony in Peterborough, New Hampshire, where I was in residence for a month, working on what would become the first chapters of *A Garlic Testament*. The English novelist had been trained as an agricultural economist at Oxford but had walked away from that career to become a novelist. Our shared interests in the crafts of writing and farming drew us together. We took to having long after-dinner conversations about agricultural economics and farming and writing in the cavernous living room of the old MacDowell farmhouse. He was intelligent, literate, soft spoken, and, as I was soon to discover from reading his first novel, a very good writer.

We kept in touch after I returned home to New Mexico. The next winter he expressed interest in coming west. Convenient: RoseMary and I were looking for a house sitter to cover for us during an extended trip. We left him with the house and offered the use of our Ford Fairmont station wagon, as he didn't have a car at the time.

When we returned from our month away, all was well, though I noted that he had put thousands of miles on the car, well beyond a normal month's driving, without mentioning it. In our absence he had arranged to become a neighbor by house-sitting another place a ten-minute walk down the river. He became a regular late afternoon visitor to the farm. I enjoyed resuming our MacDowell conversations with a somewhat wider frame of reference than that possessed by most of my neighbors. RoseMary frequently asked him to stay for dinner.

A few months into his stay, he developed what he thought was chronic fatigue syndrome. He took to reclining on the living room sofa while awaiting the almost inevitable dinner invitation. Our conversations still seemed worth the extra work he was creating in the kitchen. Eventually he asked me for a written recommendation to help him get into the Green Card lottery, for which he later obtained a winning number.

He further consolidated his position within the community by taking up with an old friend of ours who had broken up with her English husband. From his writings and from reports about visitors during our time away, we became aware of his propensity to have two girlfriends at the same time. Not unexpectedly, our friend finally broke up with him over this.

He moved to the Bay Area. During what was to be our last visit with him at a San Rafael café, he somehow concluded that RoseMary had engineered the breakup with our friend, not that it was the effect of his own double dealing. He later wrote a scathing letter to RoseMary, so vile that we read it only once before destroying it.

It was bad enough losing a friend in this way, if indeed he was really ever a friend. Others report that RoseMary and I are depicted in one of his later novels. And that the portraits are not flattering.

Chapter 37

SUBORNATION

I don't believe that Harmoni engineered our fellow garlic grower's betrayal of Ted and me: their own envy, cupidity, and paranoia led him and his wife down that path. Harmoni was quick to embrace them by offering compensation for their "research" against Ted and myself. This is acknowledged by them in a public filing with the Department of Commerce and also in the Netflix documentary. Thus did the long arm of Harmoni's legal henchmen reach into the small communities of Northern New Mexico, sowing dissent among friends and neighbors.

Had our former partner and his wife withdrawn from our cause and left it at that, a seemingly inevitable course of events would have played out. Harmoni would have been required to pay some $200 million in duties for 2015 for having failed to respond to Commerce's mandatory questionnaire. Or, more likely, Harmoni would have walked away into bankruptcy, a common escape mechanism. Duties are assessed retroactively, well after importers have distributed and sold their products, a situation that invites this kind of evasion. As of 2018, some $700 million remains in uncollected duties on garlic alone owed by importers other than Harmoni.

But no, the garlic grower and his wife went much further: they filed allegations against Ted and me in their January 2017 declarations to Commerce. Neither GDLSK or Harmoni has ever revealed how much the couple was paid for their "research." Nor, tellingly, has Commerce required Harmoni to disclose the amount of the payments.

In short, Commerce eventually accepted what could be regarded as suborned statements—false testimony induced by shady means—from our former allies as being more credible than all the filings submitted to Commerce by Ted and myself.

Chapter 38

CLOSURE?

In the fifty years I have lived in a northern New Mexico Village, I have been involved in spats, disagreements, quarrels, and fallings-out of all kinds. After a time, old friends can become like relatives: decades of common experience are inextricably woven together with personality differences, contentions over politics, attitudes toward money. But village life, in its frequent chance encounters and need for collaboration on local issues, offers an informal mechanism for reconciliation and peace-making. I have seen hatchets buried at local funerals, or at least family feuds temporarily suspended. Life must go on. If a village becomes too deeply enmeshed in feuding, it risks the fate of a legendary mountain village on Crete that was abandoned by its inhabitants because feuding had rendered it unlivable.

I would be happy to be able to sit down with my neighbor garlic grower and talk through our common history to find out where things went wrong. And to do so without accusation or blame. I considered him a friend even when he left my house on his last visit. Indeed, I long for the day when that conversation might be possible, when the deep wounds of betrayal might somehow be healed.

But what stands between us, preventing reconciliation, what stands between their village and mine, is Harmoni Spice and its several international law firms, and the lawsuits they have instigated in Los Angeles and Taos. These international corporations have created an insurmountable wall between two neighbors, between former friends. They have subverted the informal customs of reconciliation that enable small places, however contentious, to overcome internal differences and to keep functioning.

Chapter 39

SHANDONG PROVINCE

But I have got ahead of the story . . . Despite the neighbor's defection, throughout the spring of 2017 we were confident that Commerce would hold to its preliminary finding against Harmoni and that the three-year saga would finally be over. In early May, at Ruopeng's invitation, I flew to Beijing with Melinda Bateman, a garlic grower from Taos who had joined in the next Commerce review period, Administrative Review 22. With Ruopeng as our guide, we visited a garlic processing plant and its clean room, where garlic was peeled, washed, sorted, and bottled, very much like the one I had visited the year before. Parts of Shandong Province are planted horizon to horizon in garlic, the rows so close that the crop must be planted either by hand or by means of small two-wheel tractors that don't need broad rows or aisles for tractor tires. A garlic field we inspected had recently been interplanted by hand with peanuts and peppers, which would take over once the garlic was pulled. The field was covered in transparent plastic mulch, which is hand punctured when the first shoots begin to emerge in the spring.

We spent some time with a middle-aged Chinese woman perched on a small soft round stool that was strapped to her thighs in such a way she could scoot along on the ground without having to stand up. The hand tool she was harvesting garlic with consisted of a handle attached perpendicularly to a foot-long rod with a forked end. She used the tool to pry up bulbs one at a time with a quick stabbing and prying motion. She laid out the bulbs in a row, the leaves shingling over the bulbs to protect them from the sun. The bulbs were of a good size and streaked with purple. She told us how much she could harvest in a day. After converting the Chinese measures of area into hectares and then into acres, I estimated she could harvest about a quarter of an acre of densely planted garlic in a day. By contrast, with my tractor-powered undercutters, I could harvest that amount in an hour. And given the lack of rows for tractor tires, I guessed that most all garlic must be harvested by hand, not by machine. Labor is cheap. Ruopeng estimated that Chinese garlic farmers typically earn about $400 a month from their crops and for working for garlic processing plants after harvest.

During the trip, I got to know Ruopeng much better. In the intervening year his English had improved markedly. By now, he explained, he also had come to understand that American farmers could be involved in other professions, unlike Chinese farmers, who were just farmers, didn't have university educations, write books, speak other languages. For our first two days, Ruopeng gave us a tour of Beijing, including the

Forbidden City, an extensive outdoor food court, and the commercial center of the city, before we headed by train to Shijiazhuang, where he and Mary and Lucy, who worked for his export consulting firm, helped us shop.

As before, our trip included Qingdao on the coast, where we spent the better part of two days with exporter Jack Bai, who drove us out to one of his warehouses and hosted lavish meals at a variety of restaurants. Jack also warmed up and would often struggle in his limited English to utter compliments. He insisted on having one of his employees accompany us on the train back to Beijing, connect us with the car Jack had hired to take us from the station to the hotel next to the airport, and even ride with us to the hotel. Melinda and I were so lavished with gifts that we had to buy another suitcase.

Our only business conversation was very brief. Ruopeng and Jack wanted to see Harmoni charged anti-dumping duty like all other importers, so that they could re-enter the garlic export-import business. We assured them that we would stay the course, as we wanted to see the price of garlic in the US rise enough to lessen the advantage of imported garlic.

Later, I reflected on how much time Ruopeng and Jack had spent with us without ever giving the impression that they had anything better to do than spoil a couple of small New Mexico garlic growers. Mutual self-interest here, of course, but in our time together we developed a real fondness for one another, bonded in our common cause against an unscrupulous opponent.

During meals, Jack never drank because he was usually driving—except at our last dinner together at a Qingdao barbeque restaurant. Halfway through the meal he abruptly decided to down a couple of beers. A sign, I thought, that he finally considered us worthy friends and allies.

To be on the safe side, he called up a driver to drive him home and to drop us back at the hotel on the way.

Chapter 40

EL BOSQUE GARLIC FARM GOES BACK TO WASHINGTON

Shortly after our return to the States, we were astounded to learn that on June 6, Commerce had ruled against us in its final decision. The fifty-seven-page ruling cited "inconsistencies" in our answers to the mandatory questionnaire as grounds to reject the totality of our submission, and as an excuse to deny our "standing" as garlic producers. In short, paid testimony by our former partners held more weight than our own sworn statements and somehow excused Harmoni for its significant refusal to participate in the process at all. Commerce had scrutinized the declarations of my small operation far more carefully than they had Harmoni's scant filings. Commerce further closed the door by denying us an opportunity to rebut the false allegations against us. The sloppily written decision contained a major error of fact. Political intervention was written all over its pages.

Ted requested and obtained an ex parte meeting with Commerce for July. Joining Ted and Melinda and me was Suzanne Sanford of Costilla, a recent addition to the group. We met late morning in the Hoover Building on Constitution Avenue, as before. This time, there were no political appointees from the new Trump administration in attendance. Commerce staff was on the defensive from the beginning. When we and the six or seven Commerce staffers had seated, in referring to the letter from Commerce rejecting our standing, Ted began by saying, "In reference to the Gilgunn letter—"

Tom Gilgunn himself was present. He interrupted sharply, "That's not my letter, it's from the Department of Commerce."

I later thought: But you signed it, didn't you?

Melinda and Suzanne spoke about the challenges of small farmers and how keeping paper records for farmers' market sales was not required by the markets.

When my time came, I pointed out that there was something peculiar about the fact that Harmoni had not been reviewed by Commerce for thirteen years running, which even someone not familiar with the anti-dumping system would find troubling. By rejecting our request, Commerce had allowed the system to be corrupted, favoring the interests of powerful corporations.

The room was very quiet. I went on to suggest that Commerce was free to review Harmoni on its own authority, without involving us or anyone else. Commerce attorney Ed Yang, who I had met on the first visit to DC, protested that if they reviewed Harmoni, they would have to review everyone, for which they didn't have the resources. After the meeting, Ted cited a couple of cases in which Commerce had initiated reviews on its own authority. And in the months following our meeting, Commerce initiated reviews of imported aluminum sheeting and washing machines. This made Commerce's refusal to review Harmoni on its own authority appear even more troubling.

I later wondered how much time and taxpayer money had been spent in studying and responding to Harmoni's thousands of pages of filings against us. And how much more time and taxpayer money would be spent defending Commerce's decision against us in the federal Court of International Trade. Conducting a review of Harmoni's operation probably would have been cheaper. And would have ended the gestation of what has become an administrative and legal Frankenstein.

At the end of the meeting, I alerted the Commerce staff to the documentary, which would be released in early 2018, and which could eventually be seen by millions of people worldwide. The government group seemed pleased that Zero Point Zero had not contacted them. Then I slid a sack of newly harvested Russian Red garlic and a copy of the garlic book to the center of the table, hoping to end the gathering on a happier note.

We flew home that afternoon.

Chapter 41

TABOOS

One of the "inconsistencies" Commerce tried to nail us with was regarding the $50,000 I received following my withdrawal from Administrative Review 20 in March 2015. I was uncertain where the money had come from, though later Ted said it came from his personal account, as gift for enduring the stresses of the review process. Administrative Review 21 was initiated in late November, eight months after I had received the payment. In one of my Administrative Review 21 declarations to Commerce in 2016, I stated that I had received no compensation during the current review period, AR 21. Commerce, however, claimed I had received the money during the AR 21 review period. In short, it got its dates wrong.

Furthermore, in its preliminary finding for Administrative Review 21 in our favor, Commerce had ruled that our relations with exporters, financial or otherwise, were not covered by statute, and therefore the issue was moot—a point they ignored in their finding against us. Despite our many requests, they failed to investigate financial relations between Harmoni and Christopher Ranch and the other members of the California-based Fresh Garlic Producers Association.

Money is like sex: it is curtained by selective social taboos. Scandal seeks to push aside the curtains in order to expose theft, embezzlement, extortion, blackmail, and payments regarded as illicit or suspect. Publicly held corporations are legally bound to be open about money, though they have countless ways to veil transactions. Estimates of wealth of those at the very top are published annually by Forbes. From the point of view of a large corporation or law firm, the $50,000 paid to me by Ted two years before the filming of the documentary is little or nothing. From the point of view of my impoverished neighbors, it is a scandalous amount of money—as was made clear by a series of abusive emails and anonymous phone messages presumably from neighbors in my village. But spread out over three years, it was not unreasonable compensation for time spent dealing with the Department of Commerce filings. And it would cover only about two-thirds of RoseMary's annual care in the Alzheimer's facility in Santa Fe where we finally placed her.

The $50,000 had become a matter of public knowledge three years before because I had made it known—as an unexpected windfall, a reward for participating in the somewhat risky process of challenging Harmoni, which would become truly risky a year later when we were sued for racketeering. I have been cautioned about being too open about money. Perhaps I should discuss my finances only on the condition that my interlocutor also disclose the sources and amounts of his or her income. Few people, except the poor, would agree to the terms.

Most would be as reluctant to discuss their finances as to reveal their sexual practices and fantasies.

These taboos are subverted through the arts, through fiction, through TV and the movies and internet pornography, media which allow us to become voyeurs in all taboo areas, if we are so disposed. Money and sex give you power, and other people's knowledge of your dealings with money and sex gives them power over you. In scandal, all is revealed; the victim is disgraced and rendered defenseless and an object of derision. Victims are figuratively paraded naked through the streets, stripped of personal pomp and circumstance. We display our power over others in our schadenfreude. We cast our stones while forgetting that we too are by nature imperfect beings—because, inevitably, we place money in a banking system repeatedly caught bending and breaking the rules, and because, inevitably, we spend money that ends up in disreputable hands, as long as we drive cars, shop in supermarkets, use electricity generated by polluting power plants, and as long as we invest in predatory corporations that exploit their workers in sweatshops and on assembly lines in countries far removed from the reach of our labor and environmental laws, such as they are.

Chapter 42

EL BOSQUE GARLIC FARM'S FORTIETH SEASON

With the Commerce ruling against us, a new pall fell over the summer of 2017, my fortieth at farming full time. Ted challenged the ruling in the US Court of International Trade in New York, but it would be almost a year before the court would rule. The only comforting thought was that so far we had always won in court.

The summer wore on. The garlic harvest was unusually good, and we had to scramble to find covered spaces in which to stack it. We cleaned and boxed up and sold our four varieties of garlic, though we held back most of the Turban Chengdu in order to increase our planting stock. I hired a friend to make the decorative garlic bunches that RoseMary was no longer able to put together. Following garlic and shallot harvest, I planted a cover crop for the first time in years, which set off worries that tilled-in stalks of millet and sorghum might foul the close-together furrowers of the Polish garlic planter. In response I bought a flail mower in the hope it would chop stalks fine enough. In the end it chopped them up well but not fine enough. While planting garlic we had to stop frequently, lift the planter in order to shovel and rake out dams of stalks and earth. A possible solution would be to widen the wheel track of the tractor, which would allow the widening of the spaces between furrowing shovels from ten to twelve inches. I would try to do this by the next planting season.

The markets were good. The Tuesday and Saturday drill had me get up early to make coffee for interns Ben and Campbell, and later Ben and Phil, who would get up at four-thirty and leave for the market at five, driving the pickup. We made lunch and snacks the night before. They would arrive at the market around six, get set up by opening time, seven. I would leave around six-thirty, park the Volt in a parking lot with a charger, and walk to the market after picking up croissants for everyone at Sage Bakehouse. The interns did most of the selling while I caught up with old friends and signed the occasional book that someone bought. I would stay until eleven or a little later. The interns would tear down the stand and repack the pickup with the help of Scott Temple, a retired New York City Opera French horn player, who was expert at the process. I would often shop for supplies on the way home.

At the market I would occasionally cross paths with our former partner in the battle against Harmoni. We would nod stiffly to each other. I wondered whether he and his wife had thought through the effects on a small community of allying themselves with a corporate behemoth, in effect acting against the interests of all the other garlic growers at the Farmers' Market.

The summer feels complete when garlic and shallot planting is done, but the crowning event is the Dixon Studio Tour the first proper weekend in November, when we hope to sell much of the garlic remaining in storage, along with garlic arrangements and my books. That year, we also began selling a volume of RoseMary's poems for children, illustrated with wood block prints by her sister Zita. Our daughter, Kate, did the layout and arranged the printing of the book. Sales helped cover some of the expense of moving RoseMary to the Alzheimer's facility in Santa Fe. She no longer understood the meaning of the publication, but when old friends read her the poems out loud, she often broke in and finished them from memory.

In late November, Ted filed for the next period of review with Commerce, Administrative Review 23. But we also learned that Commerce had ruled against us in Administrative Review 22, which covered 2016 imports by Harmoni. In effect, Commerce moved allegations up from the previous review—contrary to statute, which declares that each Administrative Review must be separate—to Administrative Review 22, finding that because our submissions were not "credible," we didn't have standing as garlic producers. Commerce was asserting that my tax returns, depreciation records, employment records, and sales records did not count. Or worse, by implication, that they were fraudulent.

Based on some legal quibble, Commerce also rejected our submission for Administrative Review 23, which covered 2017 imports. We sensed again the heavy hand of political pressure in these actions. But then a ray of light appeared: a large Southern California company filed their own Request for Administrative Review of Harmoni.

We were no longer alone—as long as they stayed the course and weathered the inevitable attacks by Harmoni.

The hope was short lived. They later withdrew their request for review. We could only speculate that they had been bought off.

Chapter 43

HOW TO PEEL GARLIC

The Netflix series, in the end titled *Rotten*, was released in early January 2018. I invited Ted and Renate down from Taos to watch it with me in the evening, though they had roused themselves at two that morning to have a first look. The main framing device of the series was me reading passages from *A Garlic Testament* while seated on our sofa next to the fireplace. It's not easy viewing oneself on TV for an extended time. The strongest segment was of East Coast importer and processor Mingju Xu talking slowly but emotionally about how he had surreptitiously taken photos of Chinese prisoners sitting at long tables peeling garlic. He had been told that when the prisoners' fingernails gave out, they took to cleaning the cloves with their teeth. Two sellers at Chinese outdoor markets were asked where they had obtained their peeled garlic. The response of both was: "From the prison over there." Outside the prison, Xu also took photos of packing boxes inscribed with the names of subsidiaries of Harmoni and Christopher Ranch.

The episode started out well, with a clear summary of Harmoni's place in the Chinese garlic export business and our role in challenging Harmoni: a good David-and-Goliath story. But gradually it deteriorated into a personal David vs. David story. Some friends thought I came off badly for having accepted money for participating in the review process two years before, others that Xu and I were the heroes of the piece. Most thought that our former partner and his wife, who acknowledged being compensated by Harmoni, came off as being in it only for the money, and some even thought them despicable for allowing themselves to be bought off. They were Harmoni's only defenders in the episode. Christopher Ranch declined to be interviewed, a major PR blunder: they were soon savaged in social media over the prison garlic issue, about which four central California newspapers and several TV stations ran stories.

Unfortunately, the episode missed the fact that Harmoni walked away from the Commerce questionnaire, thereby subjecting itself to $200 million in anti-dumping duties for 2015. Commerce later released them from those duties when they rescinded Administrative Review 21. Nor did the episode discuss how our former partner's testimony against us gave Commerce an excuse to let Harmoni off the hook.

A few weeks after the show's release, we learned that Xu would join our side as an "interested party" in the Commerce administrative review process. It was too late in that period of review for him to request a review of Harmoni, but as an interested party, he could file comments on the proceedings.

Chapter 44

THE CONSOLATION OF THE TIMELESS

There is something bipolar about the legal machinations in distant capitals in contrast to the realities on the ground, in the earth, in the soil, of my small farm, which is where I live and where I have invested so much of my life.

Above all, it is quiet on the farm. I take the quiet for granted. After a day in the city, I crave the quiet. Sitting outside on the lawn on a warm winter afternoon, I can hear the faint sound of a car or truck passing on the highway a half mile to the south, through the leafless cottonwood trees lining the river. But the air is more likely to be dominated by the chirping of birds around the feeder hanging in front of the living room: sparrows of several kinds, juncos, chickadees, towhees, house finches, the occasional nuthatch and ring-neck dove—and rare squawking blue jays and stellar jays. There will be the caw of a raven, the yapping of a magpie, the dawn and dusk yodeling of coyotes, the bark of a distant dog, and now and then the hum of the milling machine at the cabinet maker's workshop two doors down, the occasional whine of a chainsaw in winter, a Weedwacker in summer. Then, when the human sounds are suspended for a few minutes, there rises from the river a faint wash of sound filtered through the bare branches of willows and cottonwoods, the native white noise of the valley. In the fall, in quiet afternoon moments, the plaintive honking of sandhill cranes sprinkles down from a thousand feet up. Gone for the past decade or so are the sounds of cattle, horses, goats, and sheep, at least in my neighborhood, though now and then my roosterless hens will kick up a squawking fuss among themselves.

Absent is the fossil-fuel driven thrum of traffic, often audible, even in quiet urban residential neighborhoods, at all hours. But cities are more than just noisy. They are vast three-dimensional instruction manuals. Upon entering, you are directed to Stop, Go, Wait, Walk, Don't Walk, go One Way, not another, Park here, No Parking there, No Loitering, No Smoking, No Waiting, No Texting, No Left Turn, No Right Turn on Red, No Littering, Trash, Recycle, Speed Hump, Slippery When Wet. Signs will direct you where to buy and to sell, where and how to be educated and trained and entertained, where to seek administrative and legal and political assistance, who to vote for, where to find the police, the fire department, and medical help and counseling, where and how to find companionship and sporting venues, where to eat and drink, where to rest, sleep, and where the bodies of the deceased are accepted.

A large city's menu can be overwhelming. And oppressive, as encoded in ever-present advertisements, and in the implied competitive pressures. Upon returning from the city to the village and then down our quiet lane to the farm, I can feel a sense of relief from the urban pressures. Relatively speaking, I live in the middle of nowhere, a rocky slope spotted with stunted juniper and piñon immediately above the house to the north, mirrored a half mile to the south by another such with a blackened slash from a fire set off by a lightning strike several years ago.

These slopes, these rocks, these trees, the river, they speak of nothing to the human condition. They do not instruct. They are indifferent to anything and everything I do. They will be here long after I am gone, long after we are all gone, in some form or other, certainly the rocks and boulders, probably not the trees, maybe not even the river. It's hard to think of them as witnesses, except as internalized presences. Yet their indifference can offer a sense of freedom from urban expectations and pressures. Their presence can say: do what you want, invent what you want, create everything from nothing. Start afresh. There are no limits, no boundaries. There is no obstacle to inventing yourself or reinventing yourself.

This can be the relief, the consolation of the rural, as long as you can accept the embracing indifference and see it as a positive backdrop, not as the Void, as long as you can appreciate the freedom of being in the "middle of nowhere."

Chapter 45

ANNIVERSARIES

It all comes back in the days around September 11. How we hoisted the trunks and boxes up onto the full-length roof rack of our 1966 VW camper, how we stowed suitcases and food and my handmade wooden crib for our one-year-old son inside, how we set off one October morning for New Mexico with all our belongings. Not in a panic, not even in much of a rush, because at the last minute we had become somehow reconciled with a San Francisco deeply polarized over the war in Vietnam, and over Black Panthers, Flower Children, and other liberation movements swirling around the Bay Area. The year was 1969.

Northern New Mexico offered a refuge from the big city, from urban street violence, from the rhetoric of extremists at both ends of the political spectrum, from being a target, a victim, a detainee, at a time when you couldn't avoid wondering how badly it might all end. My expatriate days in Greece and Paris and Dublin were bracketed by assassinations. I left the States to go write novels in Greece a month before the assassination of John Kennedy in Dallas in 1963. I returned five years later with my Australian wife and our newborn son in 1968, just after the assassination of Robert Kennedy in Los Angeles. By 1968, under the burden of almost irreconcilable responsibilities as a writer and as a new father, I had concluded that the global industrial systems that fed us and provided us with food and fuel and clothing and other necessities had become too complex to understand, let alone rely on. Worse, the government wanted to send every able-bodied young male off to an incomprehensible and futile war.

Less than five months after leaving San Francisco and settling into our rented house in the Embudo Valley, we planted our first garden. A year later we were bending over wooden ladder-like adobe forms and pouring mud for adobe bricks for our house, with the help of two young families of Carlisles, fellow urban refugees, while the toddlers were parked in the shade of cottonwood trees, in playpens and perambulators. By 1973, we were selling our produce first at the Taos Farmers' Market in front of the courthouse and soon after at the Santa Fe Farmers' Market in the parking lot of the Alto Street Senior Center.

The following thought now seems quaint, but at the time it didn't: If we farmed and things got really bad, then the government would give us gas ration cards and we could get to town to shop for supplies. Gas ration cards? My father had a "B" sticker on our 1939 Ford sedan, enabling him to drive to San Diego High, where he taught "industrial arts"—wood, metal, and electric shop—during World War II. As a preschooler, I stood in line with my mother while she bought sugar and other staples with her ration book. After the war, I watched her sew up muslin parcels of chocolate and coffee at the kitchen table and label them with India ink to send to her father near war-devastated Stuttgart.

But Vietnam and the social and political unrest of the sixties didn't hit in a quick series of fiery explosions on a bright clear morning, instantly changing the lives of hundreds of millions of people. It came as a succession of blows, at first seemingly isolated, only later revealing themselves to be interconnected, as they cut down an ever-widening swath of victims both obscure and prominent. In Berkeley I once ran for my life from the heavily armed TAC Squad attempting to clear Shattuck Avenue of protestors. That afternoon back in our Thirty-Sixth Avenue apartment in San Francisco, I suffered a brief nervous collapse in the bedroom, where I sat on the floor in a corner, sobbing. A few months later we were on our way to New Mexico. We crossed the border on US 60 near Quemado, in the early morning of a clear October day, after a night camping.

In struggling to comprehend the dimensions of September 11, 2001, and the ensuing wars that continue to this day, into the second decade of the century, I keep turning back to the relatively slow-motion sixties. The comparison of the two eras may seem jagged but it's not entirely useless. How did we cope then with uncertainty and fear? How did we react? Did we make the right decisions?

Two months after September 11, I flew to New York to be with friends. They were still in aftermath mode, willing to pause in their busy daily lives to retell stories of near misses, narrow escapes, heroic efforts, moments of revelation. The future had not yet emerged from the dust of still-deserted downtown

Manhattan, which reeked of drywall under layers of chemicals. I remember peering through the dusty plate glass of a closed dry cleaner at the rows of coats and dresses that would never be claimed.

Since 9/11, the future continues to darken, politically and environmentally. I wouldn't be surprised if young urban couples with children are now having the same thoughts that guided RoseMary and me to pack up our trunks and boxes and push them up onto the top of the VW camper forty-five years ago and head for the hills.

At the small, personal level, I am cheered by one small thing: at least I know how to grow our own food.

Chapter 46

MEDALLIONS

After planting and harvesting crops for over forty years, you would think a being might finally comprehend the ephemeral nature of all things. Not, alas, this one. At least not within my deep inner recesses, in the private folds of knowing.

Superficially, of course, *I know.* I've read Thucydides and Gibbon, tramped through the columns and battlements of temples and palaces reduced to marble gravel sparkling underfoot. Nothing lasts. And surely even less will last in a society that manufactures for shelf life and planned obsolescence and only occasionally constructs for all time, or most of time, or at most for a few hundred years. As a society, we are young, ignorant and foolish. We are told that the end is near. We don't know just how near, though new calculations are announced almost daily.

Privately, childishly, irrationally, I believe otherwise, denying the likelihood that our little mud house and our farm will not weather flood or landslide or drought or fire or real estate subdivision or civil disorder or radioactive cloud, and that sooner or later all will fall into ruin or be swept away or torn down, leaving at best a few little mounds. What chance is there of extending your influence much beyond your own life and that of your children, in a house built by hand, in books written, in works of art? I can remember only a handful of details about the one grandparent I personally knew. Oblivion will come within a generation or two, if not sooner. In our heartbeat biological time, living memories quickly pass into the second hand, and these in turn become distilled into a phrase, a quip, a joke, a taunt, a photograph of a face no one knows the name of, an antique curiosity, a single entry in a genealogy, perhaps a leaning granite headstone, before these small traces are swept away. All that survives of my great-grandfather is a gold-headed cane engraved with his name, plus the memory of how he as an old man would polish the brass push plate of the swinging kitchen door of the Riverside, California, house, only to be irritated whenever someone left fresh fingerprints on it.

It is unbearable to think of the seas of hope and passion and suffering that lie in the soil underfoot, and in the water, and in the very air we breathe—when our own solitary pain can so easily become intolerable. And to imagine the pain and death and destruction that every advanced society feeds on, in the remote places of the earth, in the mines, the forests, oil fields, the prisons and internment and refugee camps, and on the battlefields past and present. In order to live, we must deny the misery. Civilization is above all a state of continual denial, as is perhaps life itself, except for those who traffic directly in demise, such as the military, doctors, undertakers, butchers, executioners, policemen, journalists, all of whom work on our behalf, in our name.

History proves that nothing lasts. Or that almost nothing lasts. That bits and fragments and shards and DNA are the rule. Over the course of the next couple of hundred years, my village will wax and wane and perhaps even be abandoned in a time of extreme drought or from a meltdown at Los Alamos National Laboratory that blankets the region in radioactive dust. The river will rise and dry up and rise again, and perhaps the beavers will flood the field where we farm, rebuilding the topsoil they themselves first created millennia ago with their dams and ponds. Then, some distant day, some creature, some human rooting around in the earth, planting a garden again or a tree, may come across a thick, rough ceramic disk an inch and a half in diameter, with a hole in it, and on one side embossed letters in a script that time has rendered indecipherable, EL BOSQUE GARLIC FARM or AJOS DEL BOSQUE, the letters circling an image of a garlic bulb in bas relief. This future gardener, if human, may slip the disk into a pocket or sack or tie it with a string and hang it from his or her neck. The finder's day may pass in wonderment at what the disk might have once meant, and at who those strange people might have been, and what they once did on this land. In like manner I have pulled from my fields rusty horseshoes, a bridle bit, fragments of metal spurs, and have wondered about those who preceded me. And on the mesa above, there lie pottery shards pre-dating the Conquest by hundreds of years.

But why should we wish to be remembered, wish to become the exceptions, in the churning cycles of life that we both preside over as farmers and gardeners and are in turn presided over, within the upheaving geological cycles whose yawning eons are embodied in the very rocks that have witnessed generation upon generation of us heartbeat creatures tramping past? Why do we not simply accept the undeniable, the universal fate of the living? Why, instead, do we wish it all to go on forever, in some form or other?

Because perhaps both the comedy and tragedy of life lies in its capacity for denial, for pretending otherwise, for acting as if the universal Consequence did not apply in this or that case, for imagining something quite other. The biological world, in us, in achieving imagination, the ultimate tool, the tool of tools, has broken through the limits of all living things. We will not live forever, but we can imagine living forever, we can imagine our works living on forever, and this perhaps is enough. A lie, the cynic may argue, but what else is the imagination, even the cynic's, but a fluid or a gas that works unceasingly to dissolve or crack the nearly eternal certainty of the rocks, the stones, the boulders, and seeks to escape limitations of all kinds, and to hover over it all, for all time. Empires collapse, as this one will, but history lives on to reinvigorate what has been turned into piles of stone, in a triumph of imagination over time.

A succession of local potters has stamped out thousands of stoneware and terra cotta disks for us over the past forty years. We have tied them to thousands of garlic braids and bunches, sending them out into the world like notes in a bottle. In time the garlic is either eaten or dries up, the white husks become dusty, and the arrangement is thrown away. But now and then, I imagine, someone cuts off the disk, puts it on a windowsill or in a drawer, or ties it to the end of a closet light pull-cord. Now and then a customer at the farmers' market will hand us back one to recycle onto the next braid or bunch.

Perhaps what we are doing with these ceramic disks is sending spores out into the far distant future to prove that we once existed and labored and thought and felt. And as a test to see how far our imaginations could reach, in time and place, through whatever extreme circumstances the future might impose on the Earth, in the form of simple ceramic disks, our private little plaques like those more public ones on buildings and by the side of the road and in public parks and those left on the moon and sent into outer space bearing hieroglyphic messages saying not much more than:

HUMAN, ALIVE.

Human, alive. That's now, and will always be now.
As for Then, I'll leave it to the imagination.

Chapter 47

APOCALYPSE SHORTLY

From late spring until early summer, our main work on the farm is dealing with weeds. They like to grow twenty-four hours a day. They grow while you eat and rest and sleep. They grow especially well under the fine-spun white polyester floating row covers I lay down over new plantings of lettuce and other greens to protect the crops from nibbling flea beetles and thrips and squirrels and cottontail rabbits. Weeds are markedly fond of drip irrigation. Combine the two, row covers and drip irrigation, and you will be amazed at how well you can grow weeds.

We should be out there weeding twenty-four hours a day, not a mere four or six. I say "we" because the El Bosque Garlic Farm weeding team usually consists of me and another worker or intern or two. Up until she had trouble distinguishing weeds from crops, RoseMary was out in the field with the rest of us.

But now and then we kick back and say, "Hell with it, let 'em grow, let 'em flourish," fooling them into thinking they will be able to take over. Time to walk away from it all, catch a few plays in New York, which RoseMary and I did for three days once in mid-May, a vacation our weeds especially relished.

Or just spend a nice spring evening with friends. A notable escape some years back was a potluck dinner up the First Arroyo, the Arroyo de la Mina, in Dixon, at activist-builder-painter Hank Brusselback's new house, at a time when his wife, Gaia, had finished transitioning down from Boulder, where Gaia had taught at the university. A dozen of us Dixonistas left our weeds and evening chores in order to inspect Hank's paintings and new building projects, which included a photovoltaic array, and to gather in their living room to savor each others' garden produce and cooking. After dinner, while half the party wandered around the house and yard, the rest of us found ourselves in a circle around the sofa discussing our favorite topic, the End of the World. A few of us were reading Jared Diamond's *Collapse: How Societies Choose to Fail or Succeed.*

The conversation started innocently enough about the old days and how there were once five stores and three gas stations in Dixon, now only one and none, and the even older days when the Denver, Río Grande, and Western Chili Line narrow-gauge steam trains were still puffing up and down the Río Grande Gorge twice a day. It was common in the old days for families to make a yearly 50-mile round-trip to Española by horse and wagon for bulk supplies—not a daily car trip to Walmart. Everyone in the Embudo Valley farmed and kept livestock as a matter of basic survival. Up until the 1940s, Northern New Mexico probably resembled what Cuba became after the collapse of the Soviet Union and the end of Soviet subsidies: an intensely cultivated patchwork of organic farms—organic because trade embargoes ruled out the importation of synthetic fertilizers and herbicides and pesticides. And in pre-1950 Northern New Mexico, no one could yet afford such potions, which in most cases hadn't been invented yet. For lack of hard currency to support petroleum imports, the Cuban solution for agricultural motive power was a return to oxen.

A reflective moment fell over the living room when we all imagined a future without cheap oil and throwaway petroleum-derivative products, a world in which the automobile would become a museum curiosity. We pondered the prospect of all of us having to grow our own food. I thought about my weeds, some of which are edible. Someone who had recently read Clive Pontine's excellent *Green History of the World* reported that Pontine estimated that if the planet reverted to a pre-agricultural form of existence, it could support only about four million hunter-gatherers.

It was time for us to lighten up. With some urgency, Karen Cohen, a retired public-health physician, asked, "What will we do about coffee?" In the post-petroleum world before us, when all of us will be spending most of our days planting and harvesting and irrigating (and weeding), we'll surely need regular cups of coffee to keep us going. Coffee, like most present-day imported staples benefiting from low fuel and transportation costs, is bound to become prohibitively expensive.

Someone suggested that global warming might turn Northern New Mexico into a tropical paradise. We are, after all, at the right altitude to grow coffee. And chocolate. As we all know, chocolate is very important as another one of those enticements that awaits us at the end of a long session of weeding.

This led to more general dietary questions in situations of scarcity. "We could eat pigeons," Robert Brenden coyly suggested. Brenden is a sculptor who has spent winters in Oaxaca and Guadalajara. There followed an extended discussion of whether the pigeons living in the attic of the Dixon Presbyterian Mission should be eaten first or those in the Catholic Parish Hall attic. Someone wanted to know whether squab was still on the menus of fancy restaurants.

Robert Templeton, the Embudo Valley's expert birder, advanced the interesting fact that James Audubon not only painted birds but also liked to eat his subjects. He found juncos most tasty. Dixon conveniently abounds in the chirpy little birds during the winter. Junco pie? Junco stew? Junco frittata?

Cohen, Templeton's wife, spoke up again. "That still doesn't solve the coffee problem."

Or how to fend off the marauding hordes of Santafesinos who will look north to Dixon as the bread basket, or salad bowl, of Northern New Mexico and who, as the supermarkets begin running out of their three-day supply of salad mix, will not much care about the niceties of property rights. Fine, fine, I could hear myself saying, take what you want from the garden, but could you pull a few weeds on your way through? Pointing out that some of them were quite edible, notably lambsquarters and purslane.

Brusselback, our host, suggested that given the rapid approach of the end of the world, it would be fitting to assemble an Armageddon Cookbook. One of his apocalyptic paintings would make a fine cover. Robert Brenden, the sculptor, reported the existence of a Mexican rat poison called *La Última Cena*, or the last supper. "And there are always grasshoppers, you know.

In Oaxaca, if you want to return, you have to eat at least four of them before leaving."

Silence fell over the small gathering as we contemplated the petroleum-less years ahead and the intractable problems of how to obtain chocolate and coffee—and a regular supply of fresh croissants from Santa Fe. As for the rest, we knew we could all put up with that. The place has not been called the Independent Republic of Dixon for nothing.

People yawned and stretched, looked at their watches and cell phones. Finally our host stood.

"Coffee anyone?"

Chapter 48

THE WORLD-WIDE WEB

Laws are spider webs through which the big flies pass and the little ones get caught.
—Honoré de Balzac

The web Ted Hume and I have been entangled in is made up of the following law firms opposing us: Grunfeld, Desiderio, Lebowitz, Silverman & Klestadt LLP (New York; Washington, DC; Los Angeles; Hong Kong); Kelly Drye & Warren LLP (New York; Washington, DC; Los Angeles; Chicago; Houston; Parsippany, NJ; Stamford, CT); Winston & Strawn LLP (North America, Europe, Asia), Bardacke Allison LLP (Albuquerque); the US Department of Justice; plus Chinese law firms undertaking actions in China against our Chinese allies.

Besides the US Department of Commerce, our cases have been or are currently before the US Court of International Trade (New York City); the Federal Circuit Court in Washington, DC; the United States District Court (Central District of California, Western Division, Los Angeles); the Central District US Appellate Court (Pasadena); the New Mexico Supreme Court (Santa Fe); and New Mexico State District Court (Taos).

Firms that have carried or are currently carrying our cases include Lanza & Smith (Irvine, California); Keleher & McLeod, PA (Albuquerque); Sheehan & Sheehan PA (Albuquerque); Raiti PLLC (New York City); the Natelson Law Firm (Taos); International Technical Business Consulting (Washington, DC); and Heller & Edwards (Beverly Hills), representing fellow RICO defendants; plus those law firms in China representing our allies there. At least five attorneys (in Atlanta, DC, and Shanghai) have offered informal advice.

I have been asked a number of times whether I regret becoming involved in this labyrinth. No, because it has been a fascinating peephole into how the world works, and an education on how laws are distorted to serve the wealthy, and how the "free trade" system has been gamed to the disadvantage of domestic producers. This is the world most people enter the moment they pass through the automatic doors of a Walmart, Harbor Freight, or any number of other big box stores whose shelves are lined with imported goods.

I have also been asked what might be the connection between Trump's trade war with China (as of 2019) and our battle to have anti-dumping regulations enforced. For one reason or another, there doesn't seem to be any connection. One could argue that if anti-dumping regulations already on the books were actually enforced, instead of subverted by means of widespread gaming of the system, there would be no need for new tariffs.

Would I do it again? Yes, with one caveat. I regret only one thing: that I invited my neighbor garlic grower to participate in the fight against Harmoni. His betrayal has been among the most painful episodes of my life.

Chapter 49

ENDPAPER

In the fall of 2014, when I decided to file the first request for review, the object seemed simple: to end Harmoni's long-standing zero duty rate in the hope that the price of imported garlic would rise enough to benefit small US producers such as myself. I also hoped that it would also allow other Chinese importers, at least one of whom became an important ally, to resume trade with the US market. Attorney Ted Hume's interest was in seeing an end to the egregious abuse of the anti-dumping system that had allowed Harmoni to legally game the system.

But in what might become a reverse onion effect, our efforts became layered over first by thousands of pages of filings against Hume and myself by Harmoni's law firms and then by a string of Harmoni-led lawsuits, calling into question our constitutional right to petition the government for redress of grievances, impugning our reputations, and suborning a fellow garlic grower. In all of this, our original intention, to attempt to level the playing field between the price of imported garlic and that of domestic producers, became buried under the legal layers.

After a certain point—when Harmoni filed the racketeering suit against us—there was no pulling back, no dropping out. Nor did their increasingly extreme actions seem to be about the prospect of having to finally pay duty on their garlic imports. Were they out to teach a lesson to small a garlic producer and a sole-practitioner trade attorney? Were they hiding even more egregious business practices?

Young David, it is to be recalled, took up his sling and five stones against the Philistine Goliath, because the leader of the Israelites, Saul, was fearful of doing so himself. As of November 2018, we have lobbed some of our five stones. The Goliath still stands, armored by his international law firms. In the meantime, shades of Jarndyce & Jarndyce, the original US District Court judge in LA has left the case. She died at age fifty-two.

Chapter 50

HAPPY ENDINGS

When Ted and I started this journey, we thought it would be a simple process, completed within a year. Harmoni's business costs would be reviewed. It would be assigned a new rate of duty, which could be as high as $4.71 a kilo, or as low as zero—as it had been for the past ten years. Whatever the outcome, at least Harmoni would be reviewed.

But Harmoni managed to turn the process on its head. For the five years since our first request, they have continued to evade the review process while succeeding in re-directing Commerce's focus to my small garlic farm. Thousands of pages of allegations against my farm, and against Ted, have been filed. Legal costs on both sides have run into the tens of millions of dollars, not to speak of taxpayers' resources consumed by Commerce in studying and commenting on the voluminous filings by Harmoni, plus our extensive responses.

Five years in, nothing remains settled. As of June 2019, we are awaiting legal decisions in three courts and an administrative decision with the Department of Commerce. Of course we hope for happy endings on all fronts. And of course we try not to think of the consequences of losing any or all of our legal cases.

A happy ending of the racketeering suit against us would be for the judge of the Ninth District Court in Los Angeles to accept our motion to dismiss, effectively ending the four-year old case.

A happy ending in the Federal Circuit Court in DC would be a decision reversing the Federal Court of International Trade decision against us, which confirmed Commerce's cancelling Administrative Review 21, letting Harmoni off the hook for $200 million. If we win, and if the Department of Commerce appeals, the case could eventually end up in the Supreme Court.

On the basis of flawed procedures, we are also contesting Commerce's decision to rescind Administrative Review 22, essentially excusing Harmoni for being reviewed for the thirteenth consecutive year.

For Administrative Review 24, which covers imports for the year 2018, we are yet again requesting that Harmoni's zero duty rate be reviewed. A satisfactory outcome here would be for Commerce to accept our request and finally review Harmoni.

One of the main issues that is being argued in all of the cases is my standing as a producer of garlic, which has never been a question in the minds of my thousands of farmers' market customers and thousands of visitors to the farm during my weekly Farm Friday open house and thirty plus years of the annual Dixon Studio Tour in early November, when I often sell the last of the season's garlic.

I'm happy to report that I am likely to harvest a bumper crop of garlic for the summer of 2019. The legal and administrative machinations may drag on for years and years. But one of the consolations of farming is that while I am out in the field planting, manuring, weeding, and harvesting, physical labor has the effect of relieving me of all thought of anything else beyond the earth at my feet and the tens of thousands of plants under my care.

Yes, there are aches and pains. Yet in such labor lies an ongoing pleasure that has no need of a happy ending: each new day yields its own rewards and joys.

POSTSCRIPT

In late August 2019, Harmoni Spice agreed to dismiss the racketeering case against Ted Hume and myself, perhaps realizing that this sham attack could go nowhere. In exchange, Ted and I agreed to submit to interviews by Harmoni's attorneys, though no information obtained in them could be used against us in any legal or administrative procedures. No money changed hands. The settlement did not prevent us from requesting that the Department of Commerce continue to review Harmoni's zero duty rate. We were also free to pursue our cases contesting Commerce's decisions against us in the Federal Court of International Trade and the Federal Circuit Court in DC.

Tony Lanza, our California attorney, negotiated the settlement. He considered it a great victory for us.

So much for that particular vampire.

ABOUT THE AUTHOR

Stanley G. Crawford is a writer and a farmer. He is co-owner with his wife, RoseMary Crawford, of El Bosque Garlic Farm in Dixon, New Mexico, where they have lived since 1969. Crawford was born in 1937 and was educated at the University of Chicago and at the Sorbonne. He is the author of nine novels, including *Village*, *Log of the S.S. The Mrs. Unguentine*, *Travel Notes*, *GASCOYNE*, and *Some Instructions*, a classic satire on all the sanctimonious marriage manuals ever produced. *High Country News* describes him as one of the most original and incisive authors writing about the West today, and calls *Village* "a quiet masterpiece." He is also the author of two memoirs: *A Garlic Testament: Seasons on a Small Farm in New Mexico* and *Mayordomo: Chronicle of an Acequia in Northern New Mexico*. He has written numerous articles in various publications such as the *New York Times*, the *Los Angeles Times*, *Double Take*, and *Country Living*. For more information, please visit stanleycrawford.net

CPSIA information can be obtained
at www.ICGtesting.com
Printed in the USA
JSHW021137300323
39627JS00003B/2